Beyond Relief

Beyond Relief
Food Security in Protracted Crises

Edited by
Luca Alinovi, Günter Hemrich and Luca Russo

Intermediate Technology Publications Ltd
trading as Practical Action Publishing
Schumacher Centre for Technology and Development
Bourton on Dunsmore, Rugby,
Warwickshire CV23 9QZ, UK
www.practicalactionpublishing.org

ISBN 978 1 85339 660 1 (Practical Action Publishing)

ISBN 978 9 25105 589 2 (FAO)

A catalogue record for this book is available from the British Library.

Since 1974, Practical Action Publishing has published and disseminated books and information in support of international development work throughout the world. Practical Action Publishing (formerly ITDG Publishing) is a trading name of Intermediate Technology Publications Ltd (Company Reg. No. 1159018), the wholly owned publishing company of Intermediate Technology Development Group Ltd (working name Practical Action). Practical Action Publishing trades only in support of its parent charity objectives and any profits are covenanted back to Practical Action (Charity Reg. No. 247257, Group VAT Registration No. 880 9924 76).

Produced with financial support from the EU.

Cover image © FAO/6098/H. Null
Cover design by Mercer Design
Typeset by SJI Services
Printed by Replika Press

Contents

Preface

Over the years, it has become clear to many working in humanitarian and development fields that an increasing number of crises do not fall neatly into either of these broad categories. This is particularly true for protracted crisis situations where what were originally considered emergency situations continued over years and even decades. Indeed, it soon became clear that there was a huge policy gap and a lack of suitable frameworks to guide response and longer-term programming in these complex and volatile situations. Therefore, many leading practitioners came together in 2003 in a FAO-sponsored workshop called: 'Food Security in Complex Emergencies: building policy frameworks to address longer-term programming challenges.' Discussions at the workshop clearly pointed out the need for more evidence-based research and information regarding protracted crises.

It was therefore decided to commission a series of in-depth case studies on Somalia, Sudan and the Democratic Republic of Congo. All three contexts are notoriously difficult to document as they are extremely volatile, with complex dynamics that are often completely opaque to outside observers. The few attempts made to understand these crises have often been hindered by a lack of information and suitable frameworks, while interventions have been hampered by real danger, including open conflict, and the collapse of institutions. Not surprisingly then, interventions have often been inappropriate and based on generic approaches. And yet, the case studies in this book clearly show that these are precisely the situations where analysis and response cannot be generic and must be based on a deep understanding of the local contexts and structural factors which caused the crises in the first place.

As in many other protracted crisis situations, these case studies show that there has often been a weak link between emergency response and longer-term rehabilitation and development. In fragile states, which often exist in protracted crisis situations, entry points for intervention may not be clear as the state may be very weak or even non-existent. Emergency assistance is usually externally driven, with a risk of being inappropriate and even fuelling conflict. In addition, very few donors commit to longer term development, once the most acute phase of a crisis passes, and so root causes of conflict and crises remain untouched.

Even in these bleak situations, there are still some cases of what good practice for interventions might look like. These case studies thus attempt to document what did and what did not work in various food security interventions and draw cross-cutting lessons based on concrete examples. In particular, they look carefully at longer-term causes and at the roles of both formal and informal institutions. They also give some recommendations on what can effectively be done in situations where the state and institutions have almost totally collapsed.

Perhaps the most important message in this book is that getting away from a standard 'blueprint' approach for dealing with protracted crisis means understanding and responding to the long-term structural issues that often caused the crises in the first place. For example, issues and conflicts related to land, in particular, are often fundamental and yet extremely difficult to address, especially for external actors. Yet, the studies clearly show that securing adequate access to land is a pre-requisite for food security.

Another important lesson is that a deeper understanding of local institutions is vital for 'doing no harm' and not fuelling conflict as in the case of Somalia where due to an inadequate knowledge of clan politics some humanitarian and development agencies, eager to promote participation, worked with local groups that represented militia factions rather than households and communities (Hemrich, Russo and Alinovi, this volume). Not surprisingly, the studies also show that institutions had already begun breaking down long before conflict erupted. In fact, failed institutions were often as much a cause as an effect of protracted crisis. However, and on a more positive note, these case studies show that informal institutions can play an important role both in securing long-term food security and as entry points for conflict transformation, such as in the Sudan NMPACT case study.

The book is also full of surprising insights. These often centre upon unexpected resiliency. For example, while it is often presumed that markets collapse in the case of failed states, markets in Somalia for the most part were functioning very well. Furthermore, a buoyant currency, supported by an efficient informal remittance system, was established within two years after the total collapse of the state. Somali entrepreneurs also quickly took advantage of emerging technologies, in particular mobile phones, to overcome information gaps, fuel business and facilitate the transfer of huge sums coming from remittances.

Indeed, the resilience of the people living under such dire conditions as those in protracted crises comes to the forefront in these case studies. In almost all these cases, profound, occasionally successful, shifts in livelihoods also occur as coping strategies evolve into longer-term patterns. One of the future policy and programming challenges would then be how to support this resilience and certainly not to hamper it – as could happen with massive, but poorly planned, response which is not based on adequate analysis and understanding of local context, institutions and shifts in livelihoods.

To conclude, it is only by looking beyond the surface that we see that famines and food emergencies are essentially political in nature (see Sen, 1981; and de Waal, 1993). Only by understanding the complex political nature of protracted crises will we get away from the blueprints, stop treating them as short-term emergencies, and actually begin to deal with the root causes that hold the key to resolution.

Prabhu Pingali
Head, Agricultural Policy and Statistics Division,
Bill and Melinda Gates Foundation

References

Sen, Amartya (1981) *Poverty and Famines: An essay on entitlement and depression*, Oxford, Clarendon Press.

De Waal, A. (1993) *War and famine in Africa*. IDS Bulletin 24(4), Brighton, UK, Institute of Development Studies.

Acknowledgements

The editors would like to thank Prabhu Pingali, Margarita Flores, Michael Roberto Kenyi, Denise Melvin and Andrea Stoutland. They would also like to acknowledge the generous support of the European Union for making this book possible.

Sara Pantuliano would like to thank Caroline Gullickand and Tom Hockely, who shared relevant information, and John Plastow.

Timothy Raeymaekers would like to thank André de Groote and Léopold Mumbere from VECO (Vredeseilanden Cooperatie). He would also like to thank Omer Kambale Mirembe, François Paluku Biloko and Anselme Paluku Kitakya for their advice and support, as well as Pelo Muhindo Kyakwa, Jean-De-Dieu Kakule Kausa, Tuverson Kakule Mbakisya, Jean-Pierre Muhindo Musivirwa, Muhinda Mutokambali, Nzangi Muhindo Butonto, Crispin Mitono, Edmond Kambumbu Nguru, Gerard Muliwavyo and Rousseau Kasereka Musafiri for their dedication and hard work. His gratitude goes out to VECO RDC and its partners for their intellectual and logistical support and to Koen Hauspy for final logistical interventions.

Koen Vlassenroot would like to thank Salome Ntububa, Deo Mirindi, Alexandra Bilak for her kind hospitality in Bukavu and Els Lecoutere for her very useful comments on earlier drafts.

Andy Catley, Tim Leyland and Suzan Bishop would like to acknowledge Dr William Mogga, Bryony Jones, Andrew Bisson, Dale Hogland, Dr Simon Mwangi, Dr Thomas Taban, Sally Crafter, Bart Deemer, Steve McDowell, Silvester Okoth and Dr Piers Simpkin. They would also like to thank Luca Russo for useful feedback on the first draft and numerous NGOs who provided assistance as well as copies of their reports on livestock interventions in South Sudan.

Suzan Bishop, Andy Catley and Habiba Sheik Hassan would like to thank the following people for their valuable assistance with the chapter on livestock in Southern Somalia: Abdi Osman Haji-Abdi, Fritz Mahler, Stephanie Rousseau, Lammert Zwaagstra, Adrian Sullivan, Rose Jeptoo, Graham Farmer, Cyril Ferrand, Cindy Holleman, Consolata Ngemu, Nick Haan, Simon Narbeth, Ali-Nur Duale, Emily Nthiga, Imanol Berakoetxea, Chris Baker, Jack van Holst Pellekaan, Mohamed Dirie, Stephano Tempia, Alessandro Zanotta, Ricardo Costagli, Patrick Martin, Emmanuella Olesambu, Peter Little, Attilio Bordi,

Piers Simpkin, Calum McLean, Jerry McCann, Robert Bowen, Mohamed Aw-Dahir, Abdulwahab Sheik Mohamed, Dan Maxwell, Vittorio Cagnolatti, Abdullatif Abdi, Mario Younan, David Bell, Laura Powers, Kate Longley, Felippe Lazzarini, Paul Githumbi and Sara Reggio. Special thanks to Mr Mohamed Jama of SOCDA for giving up his and his staff's time to assist with interviewing livestock owners, traders and exporters in central Somalia and Mogadishu.

Peter Little would like to acknowledge several individuals and organizations that gave assistance in gathering the background materials for this book. They include Nick Haan, Carol King'ori, M. Aw-Dahir, Philip Steffen, Hussein Mahmoud, Riccardo Costagli, Friedrich Mahler, Andy Catley and Tim Leyland. He is also grateful to Luca Russo, Luca Alinovi and Günter Hemrich who provided initial guidance on how to approach this study and the writing of the chapter. Some of the data collection and field research for the chapter were supported by grants from the John D. and Catherine T. MacArthur Foundation and the BASIS Collaborative Research Support Program (based at the University of Wisconsin). These are gratefully acknowledged.

Tables, Boxes and Figures

Tables

Boxes

Figures

Acronyms and abbreviations

AAME	African adult male equivalent
AAP	Aide et Action pour la Paix
ARS	Area Rehabilitation Scheme
AU/IBAR	African Union/InterAfrican Bureau for Animal Resources
CAHW	community-based animal health worker
CAPE	Community-based Animal Health and Participatory Epidemiology
CFW	cash for work
CIAT	Comité International d'Appui à la Transition
COOPI	Cooperazione Internazionale
CPA	Comprehensive Peace Agreement
DEA	development-oriented emergency aid
DFID	Department for International Development
DRC	Democratic Republic of the Congo
EC	European Commisson
ECHO	European Commission's Humanitarian Aid Office
EPAG	Emergency Pastoralist Assistance Group
FARDC	Forces Armées de la République Démocratique du Congo
FAO	Food and Agriculture Organization of the United Nations
FDLR	Forces Démocratiques pour la Libération du Rwanda
FFW	food for work
FSAU	Food Security Analysis Unit
GIEWS	Global Information and Early Warning System
GoS	Government of Sudan
GREP	Global Rinderpest Eradication Programme
HAC	Humanitarian Aid Commission
IDP	internally displaced person
IFAD	International Fund for Agricultural Development
IFRC	International Federation of Red Cross and Red Crescent Societies
ILO	International Labour Organization
IMF	International Monetary Fund
IPC	Integrated Food Security and Humanitarian Phase Classification
JMC/JMM	Joint Military Commission/Joint Monitoring Mission
JPI	Jubba pump irrigated
JVA	Jubba Valley Alliance
LMAP	lower and middle Jubba agro-pastoral
LWG	Livestock Working Group
MDRP	Multi-Country Demobilization and Reintegration Program

MFC	Mechanized Farming Corporation (Act) (Sudan)
MLC	Mouvement pour la Liberation du Congo
MONUC	Mission of the United Nations in Congo
NMPACT	Nuba Mountains Programme Advancing Conflict Transformation
NRRDO	Nuba Rehabilitation, Relief and Development Organization
OCHA	Office for the Coordination of Humanitarian Affairs
ODA	official development assistance
OFDA	Office for Foreign Disaster Assistance
OIE	World Organization for Animal Health
OLS	Operation Lifeline Sudan
PACE	Pan African Programme for the Control of Epizootics
PARC	Pan African Rinderpest Campaign
PDF	Popular Defence Force
PRRO	protracted relief and recovery operation
RCD-ML	Rassemblement Congolais pour la Démocratie-Mouvement de Liberation
SAAR	Secretariat of Agriculture and Animal Resources
SACB	Somalia Aid Coordination Board
SCAHP	Somali Communities Animal Health Project
SEP	south-east pastoral
SHARP	Shabelle Agricultural Rehabilitation Programme
SIP	southern inland pastoral
SISAS	Strategy for the Implementation of Special Aid to Somalia
SJR	southern Jubba riverine
SKRPU	South Kordofan Rural Planning Unit
SPLA	Sudan Peoples Liberation Army
SPLM	Sudan People's Liberation Movement
SPM	Somali Patriotic Movement
SRRA	South Sudan Relief and Rehabilitation Association
SRRC	South Sudan Relief and Rehabilitation Commission
SYDIP	Syndicat des Intérêts des Paysans
TCE	Emergency Relief and Rehabilitation Division
TFG	Transitional Federal Government
TLU	tropical livestock unit
TNG	Transitional National Government
TPD	Tout pour la Paix et le Développement
UIC	Union of Islamic Courts
UNDP	United Nations Development Programme
UNICEF	United Nations Children's Fund
UNR/HC	United Nations Resident and Humanitarian Co-ordinator
UPC	Union de Patriotes Congolais
USAID	United States Agency for International Development
VCC	veterinary coordination committee
VSF-B	Vétérinaires sans frontières-Belgium
WFP	World Food Progamme

Contributors

Peter D. Little is Chair and Professor of the Department of Anthropology at the University of Kentucky (USA). He has written extensively on East Africa and cross-border trade, as well as on social change, development and globalization. His most recent book is *Somalia: Economy Without State*.

Sara Pantuliano is a Research Fellow with the Humanitarian Policy Group at the Overseas Development Institute, London (UK). Her work focuses on programming in conflict and post-conflict contexts. Prior to joining ODI she led UNDP Sudan's Peace Building Unit, developed and managed NMPACT and was a delegate at the IGAD Sudan peace process. She has also been a Lecturer in Conflict Analysis and Development at the University of Dar es Salaam and a consultant to the World Bank, DFID and various international NGOs.

Koen Vlassenroot is Professor of Political Science at Ghent University (Belgium) and coordinator of the Conflict Research Group. His most recent book, together with Timothy Raeymaekers, is *Conflict and Social Transformation in Eastern DR Congo*.

Timothy Raeymaekers is a Research Fellow with the Conflict Research Group in Ghent (Belgium).

Suzan Bishop is a veterinarian with a master's degree in tropical veterinary science and particular interest in community-based livestock development.

Andy Catley is a Research Director at the Feinstein International Center, Tufts University. He is based in Addis Ababa, Ethiopia and holds a PhD in epidemiology from the University of Edinburgh.

Habiba Sheikh Hassan is a Somali veterinarian who has spent her recent career working with civil society, particularly women's groups and international NGOs in southern and central Somalia.

Luca Russo works as Food Security Policy Analyst in the Food Security and Agricultural Project Service of the Agricultural Development Economics Division of FAO. Luca has over 20 years of hands on experience in agricultural development and food security issues, particularly in Africa. His experience covers work at community level as well as work with donors and international organizations. He also led major evaluations of humanitarian and food security policies and programmes. Luca holds an MSc in Agricultural Development from Wye College, University of London.

Denise Melvin is a communication officer at FAO's Agricultural Development Economics Division.

Günter Hemrich has worked as a Food Systems Economist with FAO's Agricultural and Development Economics Division, focusing on food security issues in crisis situations. He contributed as guest editor and author to a special issue of the Journal 'Disasters', which reviewed longer-term food security policy and programming challenges in complex emergencies. He also published work on the implications and challenges of natural disasters for food security, most recently for the 26th Conference of the International Association of Agricultural Economists. Günter Hemrich is currently Programme Coordinator of FAO's Economic and Social Development Department.

Luca Alinovi is a Senior Agricultural Economist at the Food and Agriculture Organization of the United Nations and Programme Manager of the EC-FAO Food Security Information for Action Programme. He has been managing programmes on food security and agriculture policy analysis in protracted crisis for over 15 years at FAO headquarters and in Eastern and Western Africa. He has published papers on food security and complex emergencies with the Accademia dei Lincei and ODI Disasters, and was guest editor for the special issue of Disasters on food security and complex emergencies. He has a PhD and an MSc from the Università di Firenze in Agricultural and Natural Resource Economics.

CHAPTER 1

Food security in protracted crisis situations: Issues and challenges

Luca Russo, Günter Hemrich, Luca Alinovi and Denise Melvin

Abstract

This chapter examines the definition of the term, protracted crisis, and how it has evolved over time. It also analyses several humanitarian and developmental frameworks and looks at their usefulness in analysing and responding to protracted food security crises. It thereby provides the background information necessary for a deeper understanding of the evidence emerging from the rest of the book.

Introduction

> Food security exists when all people, at all times, have physical and economic access to sufficient, safe and nutritious food to meet their dietary needs and food preferences for an active and healthy life.
> (FAO, 1996)

Few countries in the world can boast of food security, as defined above, for all of their citizens. In crises of a complex and protracted nature, and where states are fragile, achieving food security as described above remains a daunting challenge. While some regions and countries have made significant progress in reducing the number of chronically hungry people, others – particularly those exposed to protracted crises, political instability and fragility – have faced severe setbacks. In sub-Saharan Africa the number of undernourished people increased by 37 million between 1991 and 2002. This increase is largely due to changes in five war-torn countries, which accounted for 78 per cent of the region's total increase (FAO, 2006). Particularly dramatic was the worsening of food insecurity in the Democratic Republic of the Congo (DRC), where the number of undernourished people tripled from 12 million to 36 million, and the prevalence of undernourished increased from 31 to 72 per cent between 1991 and 2002.

Of the 39 serious food emergencies identified in 2006 by FAO's Global Information and Early Warning System (GIEWS), 25 were due to conflict and its aftermath or a combination of conflict and natural hazards. The DRC, Somalia and Sudan are among the five countries that have declared food emergencies during 15 or more of the years since 1986. In these countries, food crises are considered to be of a protracted nature.

Protracted crises situations tend to be highly volatile with dynamics that are difficult to understand, especially for outsiders. Adequate information and analysis is often lacking, and response is hampered by real danger and logistical problems. Analysing and responding to protracted crises are also hampered by the fact that they do not neatly fall into either side of the humanitarian/development divide. Appropriate partners and entry points for interventions may not be easy to identify, especially in the case of fragile states. The case studies in this book, by examining concrete evidence on what has and has not worked, shed light on how these situations could be tackled.

In order to make use of the lessons emerging from the case studies, it is necessary to review the thinking about 'protracted crises' and how it has developed over time. Furthermore, since the case studies critically examine existing conceptual and operational frameworks and their appropriateness in assessing and addressing protracted crisis situations, this chapter briefly introduces these frameworks and explains how they have evolved over time.

Defining protracted crisis

Conflict-related long-lasting humanitarian food emergencies have been conceptualized, analysed and addressed from a range of perspectives. In the 1990s, such situations were commonly categorized as 'complex emergencies'. These were defined as 'humanitarian crises in a country, region or society where there is a total or considerable breakdown of authority resulting from internal or external conflict and which requires an international response that goes beyond the mandate or capacity of any single agency and/or the ongoing United Nations country program' (OCHA, 1999). The discourse on complex emergencies brought into sharp focus the response capacity of the international community to deliver immediate humanitarian assistance in politically unstable situations.

In recent years, the term 'protracted crises' has been used to further emphasize the persistent nature of these emergencies (Schafer, 2002). Protracted crises are defined as situations in which large sections of the population face acute threats to life and livelihoods over an extended period with the state and other governance institutions failing to provide adequate levels of protection or support (Flores et al, 2005). Some authors (Flores et al, 2005; Pingali et al, 2005) also propose to expand the range of situations that can be referred to as protracted crises to include pandemics such as HIV/AIDS and macro-economic policy failure that, particularly in Africa, are compounded by weak governance, institutional failures and economic shocks. Based on Schafer (2002), the elements that may characterize protracted crises include:

- weakened or non-existent public institutions;
- weakened informal institutions;

- state control over part of a territory is challenged by a lack of resources and institutional failure;
- external legitimacy of the state contested;
- strong parallel or extra-legal economy;
- existence or high susceptibility to violence;
- forced displacement;
- deliberate exclusion of sectors of the population from enjoying basic rights;
- livelihoods are highly vulnerable to external shocks;
- existence of serious poverty;
- epidemic diseases;
- recurrent natural disasters.

The chapters in this book look at protracted crises essentially from a food security perspective, focusing on *the persistent uncertainties in people's access to food due to a range of interacting demand- and supply-side factors* (Flores et al, 2005). Indeed, the main characteristic of most protracted crises, in addition to the loss of human lives due to conflict, is the increasing level of food insecurity and hunger.

In recent years, with the increasing recognition of the political nature of many protracted crises, the terms 'fragile' or 'failed states' have been used to characterize situations where states have been unwilling or unable to deliver services, maintain legitimate political institutions and provide security to its people. Definitions of fragile states differ widely among donors and tend to be subjective due to the controversial nature of the concepts of stability, governance and democratization.[1]

There is a certain amount of overlap among the above-mentioned definitions. The same situations have been characterized, even contemporaneously, as complex emergencies, protracted crisis, fragile states or post-conflict transition situations by different actors. How situations are characterized is critical because it has significant implications for food security policy and programming. For example, characterizing them as complex emergencies brings to the forefront humanitarian issues and often leads to response led by the international community with an emphasis on emergency food assistance. Characterizing them as fragile states focuses more on developmental aspects of a state's capacity to deliver services to its citizens. This book, however, largely adopts the protracted crisis perspective, and thus focuses on longer-term issues and multiple causes at play in prolonged emergency situations as well as the options for addressing them.

Responding to protracted crises: Analytical and operational frameworks

Response operations for emergency situations usually involve short-term measures but, by the late 1980s, there was a recognized need to bring long-term considerations into the picture. This was particularly true in the case of long-lasting crises. Although there were relatively well-developed policy frameworks for humanitarian interventions (based on neutrality and saving lives in the short term) and development actions (based on promoting sustainability, participation and cost recovery), these were seen as inadequate in contexts where there was a need to support and protect people's livelihoods in extreme and volatile situations over years. There was a growing awareness that the complexities of the problems to be addressed required responses based on conceptual and operational frameworks that went beyond mainstream approaches. This led to the evolution of frameworks that sought to bridge the humanitarian/development divide.

The relief–development continuum approach

The relief–development continuum approach, proposed by the United Nations in 1991, was one attempt to start bridging the gap. Part of the rationale for such an approach was that development aid could also help communities reduce their vulnerability to the effects of natural hazards. Implicit in the continuum idea was that relief should be seen not only as a palliative but also as a springboard for recovery (Macrae and Harmer, 2004).

The relief–development continuum approach was a step toward linking short-term to long-term interventions. However, it assumed a linear progression back to normalcy after a shock or crisis, viewing relief, rehabilitation and development as sequential. It also assumed that crises (including conflict-related crises) were transitory interruptions in a state-led process of development (Macrae, 2001). However, faced with the reality of protracted crises, the relief–development continuum approach had significant conceptual limitations. In protracted crisis contexts there is rarely a distinct end to emergencies and progress towards recovery and development is often not linear. In spite of such limitations, the continuum's basic tenets of strengthening the links between short-term and long-term action remained valid.

Developmental relief approach

As the relief–development continuum approach evolved over the years, some agencies began promoting the developmental relief approach (Schafer, 2002). This approach was based on the contiguum concept, which rejected the mutually exclusive nature of relief and development but considered it essential to shift the focus of assistance from supporting people to strengthening institutions and processes (Duffield, 1998). It no longer saw people as passive recipients of assistance but saw them as active participants with existing strengths. The developmental relief approach highlighted the grey areas and blurred boundaries between relief, rehabilitation and development. From a conceptual point of view, it addressed some of the main concerns involved in operating in protracted crisis contexts (see Figure 1.1.)

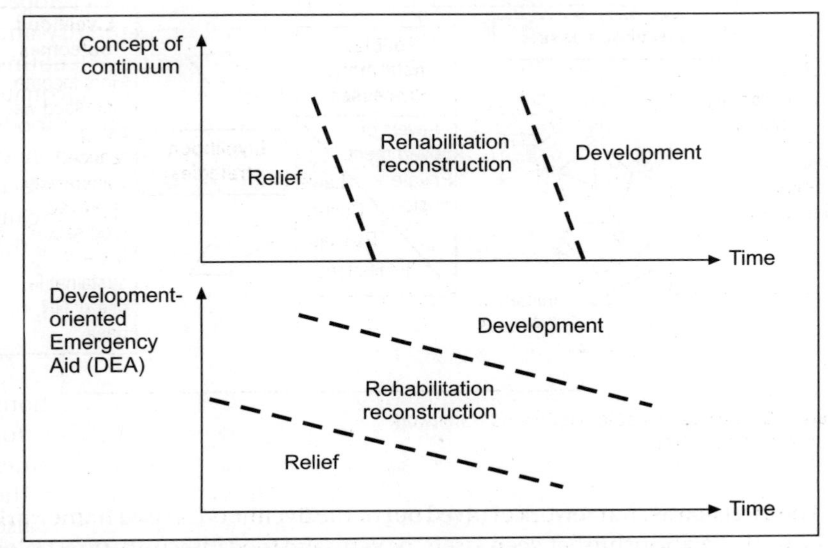

Figure 1.1 Comparison of concepts
Source: Korf (2002)

Livelihoods-based analysis frameworks

More recent thinking within the development and humanitarian communities has begun to converge around concepts of social protection and safety nets. These issues attracted particular interest when the HIV/AIDS pandemic compounded already existing structural problems and highlighted that there was a need not only for short-term relief but for long-term investments in social welfare and health systems (Harvey, 2004). In this context, two conceptual frameworks provided a platform for joint and shared planning between humanitarian and development actors: the livelihoods-based framework (with the related assets-based framework) and, especially with regard to food security, the twin-track approach.

Livelihoods-based analysis frameworks intended to capture both the main elements of people's livelihoods at a given point in time and the dynamics of livelihoods (Schafer, 2002). The basic elements of most livelihoods frameworks are assets or resources, livelihood strategies (what people do for a living) and livelihood outcomes (what goals people pursue). There are several models currently used to link these elements together. All these models draw attention to the context in which livelihoods are pursued – i.e. the policies, institutions and processes at all levels that affect people's livelihoods. One of the more comprehensive models was developed by the United Kingdom's Department for International Development (DFID) (Schafer, 2002). It has a checklist of important issues and sketches out links between them, draws attention to core influences and processes, and emphasizes the multiple interactions between different factors that affect livelihoods (see Figure 1.2).

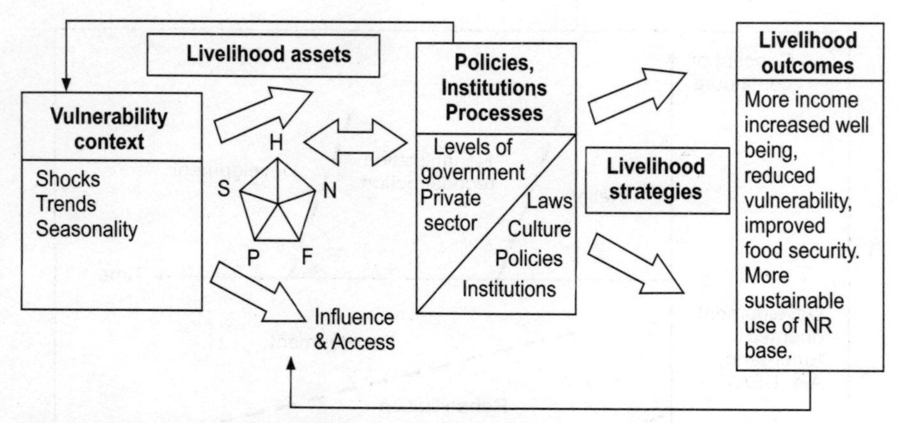

Figure 1.2 DFID sustainable livelihoods framework
Source: DFID (2007)

The assets-based framework evolved out of the livelihoods-based framework. It shows the relationship between assets, poverty and food insecurity thresholds, and recovery before and after hypothetical shocks (Barret et al, 2007). By using assets-based wealth groups and incorporating a time dimension, the model may be used to describe and monitor how households move in and out of poverty and their vulnerability over time. The framework has interesting potential for assessing the role of assets in food security and the resilience levels of different food insecure groups. It has been applied empirically in Ethiopia (Little, 2005) and Somalia (Little, this volume).

The twin-track approach

In 2002, the Rome-based UN food agencies advocated a 'twin-track' approach for hunger reduction. This framework combined investments in productive activities with targeted programmes to provide the neediest with direct and immediate access to food and other basic goods and services (FAO et al, 2002). It was later refined to include the food security dimensions of availability, access, utilization and stability, as well as related policy issues, interventions and actions (FAO, 2003). The framework provides the basis for a systematic analysis of food security indicators and responses, clearly distinguishing long-term and structural issues from those that relate to the temporary needs of vulnerable population groups. It was further refined so that it could be adapted to protracted crisis situations, with a view to linking emergency interventions with opportunities to 'rebuild resilience of food systems in periods of relative peace' (Pingali et al., 2005) (see Table 1.1).

The adapted version of the twin-track approach was designed for conducting both needs analysis and developing responses consistent with a rehabilitation or development perspective. Under track one, examples of response include improving the supply of food to the most vulnerable, reproducing locally

Table 1.1 The twin-track approach and the dimensions of food security

	Availability	*Access*	*Stability*
Track one: Rural development/ productivity enhancement	Enhancing food supply to the most vulnerable Improving rural food production, especially of small-scale farmers Investing in rural infrastructure Investing in rural markets Revitalization of livestock sector Resource rehabilitation and conservation Enhancing income and other entitlements to food	Re-establishing rural institutions Enhancing access to assets Ensuring access to land Reviving rural financial systems Strengthening the labour market and management Mechanisms to ensure safe food Social rehabilitation programmes Transfers: food/cash-based Asset redistribution	Diversifying agriculture and employment Monitoring food security and vulnerability Dealing with the structural causes of food insecurity Reintegrating refugees and displaced people Developing risk analysis Reviving access to credit system and saving mechanisms Re-establishing social safety nets Monitoring immediate vulnerability and intervention impact
Track two: Direct and immediate access to food	Food aid Seed/input relief Restocking livestock capital Enabling market revival	Social relief, rehabilitation programmes Nutrition intervention programmes	Peace-building efforts

Source: Pingali et al (2005)

improved seeds, enhancing income and other entitlements to food, re-establishing rural institutions, reintegrating refugees and displaced persons, and reviving access to credit and savings mechanisms. Under track two, the critical actions include re-establishing markets, providing food aid, cash transfers and social relief and rehabilitation programmes, and contributing to peace-building efforts (Flores, 2007). In practice, the twin-track approach ensures that the multidimensional dimensions of food security are properly addressed and that long-term and short-term food security problems are brought into the same framework.

Taking stock of concepts and frameworks in practice

The case studies in this book show how these existing paradigms and frameworks shaped, sometimes adversely, both analysis and response. In the DRC, for example, most interventions were essentially based on the relief–development continuum approach, which sees conflict as a temporary interruption to state-led development. Little attention was thus paid to resolving the root causes of conflict and food insecurity such as land access and tenure, which are critical issues not only in the DRC, but in conflicts in many parts of the world, with huge ramifications for food security.

The relief–development continuum approach has also shown considerable conceptual limitations since, in protracted crises, there is rarely a distinct end to emergency. Indeed, even in conflict situations, islands of relative peace may exist, as is currently the case in Sudan, DRC and Somalia, which are often characterized as *no peace no war* situations.

The Sudan case study on livestock is a good example of a practical application of the developmental relief approach adopted by some agencies. For example, the Operation Lifeline Sudan (OLS) Livestock Programme not only successfully controlled rinderpest – a cattle disease of major local and international importance – but also developed a primary animal health system based largely on developmental rather than relief principles (Tunbridge, 2005). The key to the success of the programme was that 'rather than regarding local people as recipients of relief aid designed and delivered by outsiders, the developmental approach used community participation and cost recovery to lay a foundation for the long-term provision of livestock services' (Catley et al, this volume).

However, the authors recognize that 'the developmental approaches initiated by UNICEF in the livestock programme would not have occurred without considerable flexibility on the part of relief donors, (Catley et al, this volume). The operational implications of this approach need still to be worked out – particularly in contexts where the approaches promoted by donors and agencies and related intervention tools tend to be rigidly divided between the humanitarian and development camps.

The Somalia livestock study also makes the case that a livelihoods approach would have made interventions in Southern Somalia more effective. For example, few if any livestock interventions were based on a livelihoods analysis that disaggregated livestock assets by wealth or gender. Consequently, interventions were often targeted not at vulnerable pastoralists but at the livestock business as a whole (Catley et al, this volume).

Indeed the livelihoods framework is increasingly used as a tool to reach across the humanitarian–development divide. Alternatively, such an all-inclusive framework underplays food security-related issues and runs the risk of being inappropriate in supporting government services that are organized along sectoral ministries.

The twin-track approach, with its special emphasis on food security, was acknowledged by several authors of this book as being particularly suitable for ex-post analysis. Pantuliano, in her chapter on Sudan comments that the twin-track approach with its 'assumption that food emergencies are social and political constructions, is consistent with the thinking that underpinned the Nuba Mountains Programme Advancing Conflict Transformation (NMPACT) initiative' (Pantuliano, this volume). Indeed NMPACT, with 'its emphasis on local capacity-building, sustainability and protection of livelihoods, delivered in its own terms and in line with the twin-track approach, successfully facilitated a collective response that buttressed the stability of the food system' (Pantuliano, this volume).

The twin-track framework is useful for identifying strengths and gaps in both long- and short-term interventions and, used as a tool for ex-post analysis, helps answer the question 'could we have done better?' (Flores, 2007). However, it is not clear how the twin-track approach can be applied in planning, prioritizing and implementing responses, particularly in protracted crises situations. It also has been noted that the twin-track framework does not adequately capture the institutional context, which is often key for planning effective interventions.

The chapters in this book look critically at the existing conceptual and operational frameworks for designing responses and analysing food security in protracted crisis situations. A one-size-fits-all framework has not yet been – and probably should not be – devised. Instead, the challenge is to integrate one or more frameworks as appropriate. The mainstream livelihoods-based frameworks tend to be weak on food security, a shortcoming that could be addressed by integrating them with the twin-track model. That model would, in turn, benefit from the institutional and dynamic perspective brought to it by the livelihoods frameworks.

Making frameworks operational and linking them to the programming and decision-making processes remains a challenge. Indeed, despite the complexity of the situations and even when adequate analysis is available, responses to food security crises have tended to be as simplistic as providing free seeds and tools.

The case studies in this book therefore aim at providing evidence-based analysis of operational and conceptual challenges, cross-cutting issues, related trade-offs and potential experience-based options for increasing food security in protracted crisis. The conclusion chapter deals in particular with the challenges of and implications for expanding country-specific lessons into a more general and forward looking framework. Furthermore, it also suggests possible options for rethinking aid delivery mechanisms in protracted crises contexts. Ultimately, the book advocates for a shift in approaches toward protracted crises, encouraging flexibility and allowing for local contexts and evidence to be taken into account when establishing policy.

References

Barrett, C., Little, P. and Carter, M. (eds) (2007) 'Understanding and reducing persistent poverty in Africa', *Journal of Development Studies* 42 (2): 167–177.

DFID (Department for International Development) (2007) Sustainable livelihoods guidance sheets, www.livelihoods.org/info/guidance_sheets_rtfs/ Sect2.rtf. [accessed 23 September 2007]

FAO (Food and Agriculture Organization of the United Nations) (1996) *Rome Declaration on World Food Security and World Food Summit Plan of Action*, World Food Summit, 13–17 November 1996, Rome.

FAO (2006) *The State of Food Insecurity in the World 2006*, FAO, Rome.

Flores, M. (2007) 'Responding to food insecurity: Could we have done it better?'. In A. Pain and J. Sutton (eds), *Reconstructing Agriculture in Afghanistan*, Practical Action Publishing, U.K. and FAO, Rome.

Korf, B. (2002) 'Challenging the continuum. The concept of development-oriented emergency aid', paper presented at the Dialogue Workshop Poverty Alleviation, Youth and Conflict Transformation in Jaffna, 28 January, Jaffna (Sri Lanka).

Little, P. (2005) 'Unofficial trade when states are weak: The case of cross-border commerce in the Horn of Africa', Research Paper No. 2005/13, World Institute for Development Economics Research, United Nations University, Helsinki.

Macrae, J. (2001) *Aiding Recovery? The Crisis of Aid in Chronic Political Emergencies*, Zed Books, London.

Macrae, J. and Harmer, A. (2004) 'Beyond the continuum: An overview of the changing role of aid policy in protracted crises', Research Briefing Paper 16, Humanitarian Research Group, Overseas Development Institute, London.

OCHA (Office for the Coordination of Humanitarian Affairs) (1999) *Orientation Handbook on Complex Emergencies*, Geneva [available at http://www.reliefweb.int/library/documents/ocha__orientation__handbook_on__.htm].

OECD (Organisation for Economic Co-ordination and Development) (2006) *Whole-of-Government Approaches to Fragile States*, Paris [available at https://www.oecd.org/dataoecd/15/24/37826256.pdf].

OECD/Development Assistance Committee (DAC) (2005) *Chair's summary: Senior level forum on development effectiveness in fragile states*, 13–14 January 2005, London.

Pingali, P., Alinovi, L. and Sutton, J. (2005) 'Food security in complex emergencies: enhancing food system resilience', *Disasters* 29 (s1): S5–S24.

Schafer, J. (2002) 'Supporting livelihoods in situations of chronic conflict and political instability: Overview of conceptual issues', Humanitarian Policy Group, Overseas Development Institute, Brighton, UK.

United Nations (2006) *Progress Report on the Prevention of Armed Conflict*, Report of the Secretary-General (A/60/891), General Assembly, Sixtieth Session, Agenda Item 12, Prevention of Armed Conflict, New York.

World Bank (2005) 'Fragile states – good practice in Country Assistance Strategies', IDA/R2005-0252, Report No. 34790, Washington DC.

PART I

Case Studies from Sudan

CHAPTER 2
Crisis and food security profile: Sudan

Luca Russo

Abstract

This chapter provides a brief background on the nature of the Sudan conflict and its effects on food security, and looks at the main actors and responses in the food security sector. It essentially covers the period prior to the January 2005 signing of the Comprehensive Peace Agreement (CPA) between the Government of Sudan (GoS) and the Sudan People's Liberation Movement (SPLM). At the time of writing, it was still too early to evaluate the effects of that peace agreement on institutions and food security, and other recent conflicts – notably Darfur – are covered only marginally.

Context

Sudan is the largest country in Africa, with a population of 30 million (7–8 million living in the south). It is a least-developed country with very poor socio-economic indicators: in 2005 it ranked 141st on the Human Development Index of the United Nations Development Programme (UNDP); life expectancy at birth is 56 years; adult illiteracy stands at 42 per cent, and 17 per cent of the children under the age of 5 are underweight (UNDP, 2005a). However, global statistics hide the tremendous socio-economic differences between the different parts of the country and in particular between north and south. It is important to recall that present-day Sudan came about through colonial interventions: the Anglo-Egyptian condominium discouraged separate administration for the north and south. Thus, the country is still highly divided in social and economic terms; while identity in northern Sudan is mostly (but not exclusively) forged around Arab-Islamic lines, southern identity is forged around 'African' lines, with continuous resistance to Arab-Islamic assimilation (Deng, 2002).

Since Sudanese independence in 1956, there has been a virtually uninterrupted series of conflicts between the Khartoum-based central government and rebel movements in the south and in other parts of the country, with a period of relative peace only between 1972 and 1983. The costs of conflicts have been heavy, particularly for the civilian population. Between 1983 and 2005 about 2 million people died as a result of the conflict and 6 million were uprooted from their homes. The various causes that explain such a protracted conflict go beyond the north/south divide; some are explicit and others underlying. They include conflicts over the exploitation of natural resources (land, water, oil) and the unequal distribution of public investments and resources. In 1980 access to primary education in the south was four times lower than in the north; the difference in access to university education was even more dramatic, with one place for every 3,500 residents in the north and only one for every 200,000 in the south (Deng, 2002). This was attributable in part to the regional authorities' lack of capacity to raise revenue but also to the failure of the GoS to comply with the resource transfer obligation required by the 1972 Addis Ababa agreement. Lack of development in the south and other regions was the major factor triggering the north/south conflict and has fuelled other conflicts, notably in Darfur.

The conflict between the north and south has been further complicated by the emergence in the south of various rebel movements fighting each other and often siding with the Khartoum government (note the role of Kerubino Dinka militia in the Bahr el Ghazal crisis of 1998). The effects of such conflicts, particularly of endogenous[1] counterinsurgency warfare, have been devastating for local institutions and social capital (Deng, 2002). In addition, neighbouring countries have repeatedly used the Sudan conflict to address internal and external problems (Prendergast and Mozersky, 2004); examples include the Ethiopia Mengistu regime providing support to the Sudan People's Liberation Army (SPLA) and the Ugandan Lord's Resistance Army's operations in southern Sudan following the signing of the CPA. Conflicts in Sudan have been characterized by phases that oscillate between deterioration, escalation, acuteness and de-escalation, with areas under relative peace and areas in conflict, and areas where control over territory on the part of the central government has been limited, particularly in the south and more recently in Darfur.

The peace process between north and south represents a first important step towards the establishment of a situation suitable for a minimum of social and economic recovery. The CPA envisages the creation of a Government of National Unity that includes 25 states, with a semi-autonomous Government of South Sudan comprising 10 of those states; three transitional areas (South Kordofan/Nuba Mountains, Blue Nile and Abyei) are the object of specific protocols. The peace process remains fragile and it is too early to project scenarios about, for example, the role of other parties and stakeholders not represented in the agreement, institutional organization and capacities, civil society participation, local factional conflicts, regional conflicts, population movements and sharing and ownership of resources.

Effect of conflict on poverty, famine and food insecurity

There is a dire lack of statistics disaggregated at the regional level on the dramatic poverty situation in the areas affected by conflict (see Table 2.1). A study (NSCSE, 2004) estimates gross income for the South Sudan region to be less than US$90 per year, about four times lower than the rest of Sudan – making it one of the poorest regions of the world.[2] Furthermore, the net enrolment ratio in primary school in South Sudan (20 per cent) is the worst in the world; the region also has high rates of infant mortality (150 per 1,000 live births) and under 5 mortality (250 per 1,000 children). In some parts of northern Sudan (Red Sea and Darfur) and the transition areas, social indicators are similar to those of southern Sudan. The rural poverty rate for all those areas is over 60 per cent (World Bank, 2003).

Widespread food insecurity is one prominent feature of Sudan's protracted crisis. Despite the country's economic and agricultural potential, over the last 10 years between 1.5 and 3 million people per year have required some form of food aid (see Table 2.2). Food aid needs assessments have highlighted regional differences and year-to-year variations attributable to the various regional crises that have affected the country, such as the Darfur crisis in 2005 (see Table 2.3).

Food insecurity remains essentially a rural phenomenon linked to the fragility of rural livelihoods. In northern Sudan, agriculture (which in 2001 represented 39 per cent of GDP (World Bank, 2003)) is characterized by four categories of farming systems: irrigated, semi-mechanized, rainfed traditional and livestock. The highest levels of poverty and food insecurity are recorded among traditional rainfed farmers and pastoralists (World Bank, 2003). In the

Table 2.1 Sudan and Millennium Development Goal 1 (Eradicate extreme poverty and hunger)

Sudan	*Current situation (%)*	*Target for 2015 (%)*
Proportion of undernourished population	26 (2000)	16
Prevalence of child malnutrition (weight for children under 5)	35 (2001)	16
Prevalence of acute child malnutrition (weight for height, children under 5)	16 (2000)	8
Southern Sudan		
Proportion of population below US$1 per day	90 (2003)	45
Prevalence of child malnutrition (weight for age, children under 5)	48 (2001)	24
Prevalence of acute child malnutrition (weight for height, children under 5)	22(2000)	11
Proportion of population facing food deficit	23 (2003)	11

Source: UNDP, 2005b

Table 2.2 Food aid deliveries to Sudan in (metric tonnes, thousands) (cereal equivalents)

1996	*1997*	*1998*	*1999*	*2000*	*2001*	*2002*	*2003*	*2004*	*2005*
108.3	114.4	201	293.5	182.4	202.8	149.4	256.1	388.8	857

Source: WFP, 2006

Table 2.3 Total number of vulnerable people in Sudan in 2005, by type and region (thousand)

Vulnerable group	*Darfur*	*East*	*South*	*Three areas*	*Other areas*	*Total*	*Percent*
IDPs	1550	75	350	95	95	2165	32
Returnees	0	0	500	395		895	13
Highly vulnerable residents	1450	245	1250	300	105	3350	50
Contingency	0	0	0	0	300	300	4
Total	3000	320	2100	790	500	6710	100

Source: WFP, 2006

south, agriculture characterized by subsistence farming, shifting cultivation and livestock and fisheries production (see Box 2.1) is the only basis for the livelihoods of over 90 per cent of the population and its economic importance has increased since the start of the conflict. The most food insecure regions or states, the traditional recipients of food aid, are: Kassala, North Darfur, North Kordofan and Red Sea in the northern Sudan; and Jonglei, Bahr el Ghazal and Upper Nile in the south. The three transitional areas (South Kordofan/ Nuba Mountains, Blue Nile and Abyei) are also highly food insecure and have received substantial amounts of food aid.

Contributing factors

Insecurity and civil conflicts are the main causes of food insecurity, particularly in southern Sudan. WFP emergency operation in 2003 had a beneficiary caseload of 1.54 million for the southern sector, all categorized as war-affected.[3]

Box 2.1 Food Economy Zones and food security

The *Flood Plains Economy Zone* is the most important in terms population and extension. It comprises areas prone to seasonal flooding, with agro-pastoralism as the dominant production system. Livestock production relies on a system of transhuman pastoralism.

The *Nile Corridor Food Economy Zone* also centres its economy around livestock, but is distinct from the Flood Plains largely as a result of the effects of the conflict(s) – which include displacement and loss of markets and assets (with increased importance of fishing and wild foods).

In the *Ironstone Plateau Food Economy Zone*, livestock cannot be reared due to tsetse flies; crops are the most significant component of agriculture (the zone is normally self-sufficient or produces surplus), integrated with wild foods and fishing.

The *Green Belt Food Economy Zone* is the traditional food surplus area of Sudan. In a normal year, self-production of food represents up to 85 per cent of household consumption.

The *Hills and Mountains Food Economy Zone* covers a large part of eastern Equatoria and is characterized by a variety of agro-ecological conditions. Livestock and wild foods comprise agricultural production.

The *Arid Food Economy Zone* represents a small area of East Equatoria bordering Kenya. Livestock rearing is the main basis for livelihoods; this is reflected in household consumption patterns, with livestock products (meat, blood and milk) representing over 70 per cent of household consumption.

Source: Save the Children-UK (1998)

The war and counterinsurgency warfare have had a direct impact on food security through:

- damage to the social and economic fabric and related entitlements;
- destruction of the already scarce infrastructure;
- mass displacement of the population;
- progressive weakening of local institutions providing required services;
- the role played by the warring factions in controlling international assistance and food relief (which is perceived as an instrument for gaining legitimacy among the population or as a means to support the war efforts)[4] and in impeding the access of operations (particularly relief operations) to areas considered 'hostile';
- changes in the food systems: for instance in the Nuba Mountains farmers living in the SPLM-controlled areas could not farm the land in the fertile plains as in the past,[5] while the traditional exchanges with the northern markets and the Bagara (Arab pastoralists) was curtailed.

In addition to the effects of the conflicts, other compounding factors have had significant effects on the food security situation:

- Natural disasters are recurrent in Sudan and the droughts of 1983–1984, 1997–1998 and 2000–2001 displaced large parts of the population and caused high livestock mortality.
- Inappropriate policies, already in place prior to the conflict, have meant that little attention is paid to development of the smallholder farmers sector. The policy focus has been on large-scale mechanized agriculture and irrigation development,[6] compounding a dramatic imbalance in resource distribution. Until the signing of the CPA, the authorities in both northern and southern Sudan ran a war economy in which food security did not receive the attention it required.
- Lack of infrastructure has had a negative impact on food security, for example by limiting the marketing possibilities for moving food from surplus to food-deficit areas.
- Insecurity in neighbouring countries has led to a large influx of refugees. The refugees, coupled with an estimated 3.5 million internally displaced persons (IDPs) scattered throughout the country, has put a further strain on already meagre socio-economic and environmental resources.

Major players: Mandates, approaches and responses

The following is a brief summary of actors that have played important roles in relation to food security in the crisis context of Sudan, with descriptions of their responses and some preliminary lessons.

Role of the international community

Until the late 1980s, Sudan received significant official development assistance (ODA), which peaked in 1985 at $1,900 million. The situation prevailing in the country brought about a radical shift in the support provided by the international community for humanitarian objectives: by 1996, more than 80 per cent of donors' resources were directed to relief and emergency operations, leaving less than 20 per cent for development assistance, which dropped to $100 million (Lehtinen, 2001). Within such a humanitarian framework, the international community addressed the effects of the conflict(s) in Sudan to the extent that it took over many of the functions normally performed by government and by 1998 OLS 'became *de facto* the Government' (Deng, 1999).

Under the humanitarian umbrella, the country was subdivided into two sectors, north (areas controlled by GoS) and south (areas controlled by the SPLM), which were allocated equal levels of resources. OLS, established in 1989 through a tripartite agreement between the GoS, the UN and SPLM following a devastating famine, included all the major UN agencies and several NGOs. It has been the main coordination mechanism for the delivery of humanitarian assistance and has reached approximately 4 million people per year. Its mandate has been essentially humanitarian (based on neutrality and impartiality) and guided by the International Federation of Red Cross and Red Crescent Societies (IFRC)/NGO code of conduct in disaster and relief, which states that the 'humanitarian imperative comes first'.

The core of humanitarian interventions during the conflict has been the distribution of emergency food assistance: it represented over 50 per cent of the total transfers made under the OLS framework, followed by health and nutrition, agriculture and food security, and water and sanitation. Food assistance during the conflict has been delivered mostly by WFP through its emergency programme category (EMOP)[7] and, within this context, food aid has been used mostly to address immediate food gaps, without building assets. However, since 2003, WFP has allocated a small percentage (10 per cent) of its food to activities with development objectives, on a pilot basis.

Another major food security initiative is the agricultural rehabilitation programme coordinated by the Food and Agriculture Organization of the United Nations (FAO). The bulk of activities undertaken by FAO (essentially through NGOs) and other stakeholders in the agriculture sector have consisted in the provision of seeds and tools to support agricultural production by vulnerable groups and have been aimed at temporarily reducing household food gaps, without reinforcing livelihoods in any systematic way.

Food security information and related analysis in Sudan have also been dominated by the humanitarian agenda. In fact, in the case of food security activities, the information generated and analysed has been mostly limited to estimating food and production gaps, following IDP movements and performing some broad-range analyses of the basic livelihoods systems (FAO, 2003). This has allowed for the design of a very limited range of responses with an emphasis on the strengthening of the 'supply' rather than the 'demand' side of existing food systems. There are notable information gaps in qualitative analyses (for example, the politics of food during civil war, social networks responses, the effects of activities beyond the mere delivery of inputs) and quantitative data (for example, beneficiaries, allocation of resources, nutritional status). Filling these gaps would support the preparation of responses with medium- to long-term perspectives; such responses could also address some of the structural causes of the conflict(s) and improve the quality and relevance of the relief interventions.

Most OLS responses have been of a humanitarian/relief nature, in accordance with its mandate. They have tended to be planned and conceived with a short-term perspective, using externally-driven planning mechanisms, with a view to producing quick and visible results on the ground. Conflict resolution issues have not been addressed, though in recent years, the OLS agencies have recognized that 'humanitarian initiatives' must extend beyond life-saving activities into building resilience and capacities for recovery (UNICEF, 2004).

Not all donors and agencies, however, have operated according to a humanitarian mandate. Several NGOs did not join the OLS coordination since they felt that the humanitarian mandate and the conditions attached to it would limit their freedom to support the people of southern Sudan. Some key donors intervened or withdrew their support to Sudan in accordance with their own policy priorities. For instance, the United States Agency for International Development (USAID) took a clear stance in the conflict because of the US political agenda.[8] It concentrated its support in southern Sudan to areas controlled by the opposition, supporting a number of responses in collaboration with the SPLM that were not of a purely relief nature nor based on neutrality principles.

Some activities undertaken under the 'emergency umbrella' have clear long-term 'developmental/recovery perspectives', even though these are often not made explicit. The support OLS agencies provide to the livestock sector under FAO coordination (described in Chapter 4) is a good example of a 'humanitarian' response with longer-term perspectives. It is different from other emergency interventions in that: (a) its coordination mechanisms are stronger than for the agricultural sector; (b) its 'hardware' component (in this case, medicines) is rather small (approximately 20 per cent) with respect to its 'software' component (capacity-building and organizational activities);[9] (c) it pays greater attention to sustainability issues, and has implemented a cost recovery strategy; and (d) it engages in policy formulation in conjunction with local institutions. Another example is the NMPACT Programme (the subject of Chapter 3), which has promoted coordinated efforts based on priorities determined by the Nuba, with a mix of short-term and long-term measures and with a focus on promoting a sustained process of conflict transformation.

With transition to peace, substantial changes are expected in the typology of external responses. However, the 2006 UN work plan for Sudan still envisages a US $1.5 billion requirement for the humanitarian sector, while budgeting only US $210 million for recovery and rehabilitation projects (US $33 million for food security interventions).

The national/regional and local institutions

In the Sudanese context, most of the 'official' responses to humanitarian crisis have been undertaken by external agencies. This can be explained in part by the dearth of qualified personnel, the weaknesses of local institutions involved in and affected by the civil war and the reluctance of most donors (with the exception of the humanitarian branch) to work with rebel movement administration.

The two chapters on Sudan and the preliminary review for this book (Russo, 2005) provide clear indications on the current and potential role of both formal and informal local institutions in food security and conflict resolution and transformation. Community-based mechanisms for handling food crises and the adaptation of livelihoods and food systems to changing circumstances have not only provided interesting long-term food security strategies, they have also proven to be effective in mitigating the effects of the crisis. For instance Harragin (1998) noted that in the 1998 Bahr-El-Ghazal crisis, vulnerable individuals were defined by the Dinka as 'those without an adequate kinship structure around them'. The households less affected by famine were those belonging to big groups with diversified livelihoods and coping strategies, irrespective of the level of individual vulnerability. Harragin (1998) and Deng (1999; 2003) describe a use of food aid by clans/kinship structures that rejected the targeting mechanism promoted by donors – which aimed to reach those the outsiders perceived as 'most vulnerable' (for example, households headed by women and IDPs) – and instead employed a redistribution mechanism within the community. Food aid was perceived as a 'common good' to be used to strengthen long-term kinship ties rather than as a means for addressing the short-term problems of certain individuals.

One perspective is to view these communities as pursuing long-term strategies based on the strengthening of social networks, which contrast in some ways with the short-term perspectives of the external assistance they receive. External agencies have tended to ignore or even resist (as in the case of the kinship-based food aid redistribution) such locally based safety net mechanisms. Social protection mechanisms based on kinship structures have been eroded by the severity and duration of the conflicts and the seriousness of asset depletion (Deng, 1999); the 1998 famine was identified by some Dinka groups as 'the famine of breaking relationships' that led to social entitlement failure.

The only significant formal local institution that has operated during the conflict in the areas of southern Sudan under the control of the SPLA has been its civil administration body, the SPLM. The main structure the SPLM put in place for agriculture and food security was the Secretariat of Agriculture and Animal Resources (SAAR), while relief operations were delegated to the South Sudan Relief and Rehabilitation Association (SRRA), later a Commission (SRRC). However, most of these administrative bodies were not transformed into effective institutions and failed to deliver the expected services, essentially because of lack of support by the international community, war-based priorities and lack of capacity. Within such a framework their inputs in terms of food security responses were focused primarily on short-term measures related to food aid needs assessment and food aid distribution undertaken through the SRRC, although some agriculture-related activities with a short-term perspective have also been promoted. It was only in 2005 that SPLM development institutions started to take a more prominent role in the negotiations with donors and in the development of specific policies.[10] Within the framework of the peace process, donors became more willing to support institutions that had been identified with the rebel movement.

The chapters

The discussion above suggests that Sudan presents many of the features characteristic of a protracted crisis situation, such as the weakened state of both public and informal institutions; challenges to central state control over part of the territory; serious levels of poverty; high susceptibility to violence coupled with forced displacement and deliberate exclusion of sectors of the population from assistance; high frequency of natural disasters; and extremely vulnerable livelihoods. In such a context, humanitarian assistance has been translated into externally driven, repeated sets of short-term responses. These responses most certainly contributed to saving lives and in some cases to protecting livelihoods. However, given their nature, they have not been able to address the structural problems of food insecurity, nor to contribute to conflict resolution or transformation.

The two chapters in this section, however, demonstrate alternative ways of responding to crises. Chapter 3, with a geographic focus, examines the complex dynamics in the Nuba Mountains and explains how an innovative participatory mechanism blending conflict transformation with humanitarian aid contributed to longer-term stability. Chapter 4, with a sector focus, looks at interventions in the livestock/pastoralist sector in southern Sudan and identifies the factors that can contribute to increasing the long-term relevance of sector-specific interventions during emergencies. A number of conditions and caveats are necessary for these approaches to work; the two chapters provide ample evidence of these. Together the chapters illustrate how, within a humanitarian context and using assistance modalities that are characteristic of emergency interventions, it is possible to promote interventions with a long-term perspective that address some of the structural causes of the conflict and food insecurity.

Timeline of conflict in Sudan

1899–1955 Sudan under joint British–Egyptian rule.

1956 Sudan gains independence.

1962 Civil war begins in southern Sudan, led by the Anya Nya movement.

1969 Jafar Numayri leads the 'May Revolution' military coup.

1972 Under the Addis Ababa peace agreement between the Government and the Anya Nya, the South becomes a self-governing region.

1983 President Numayri declares the introduction of Sharia (Islamic law).

1983 Civil war breaks out again in southern Sudan, involving government forces and the SPLA led by John Garang.

1989 National Salvation Revolution takes over in military coup. Drought and famine in various parts of the country. OLS begins humanitarian assistance.

1991 SPLA loses Ethiopia's support with the fall of the Mengistu regime. Major split within SPLA along ethnic lines.

1993 Revolution Command Council dissolved after Omar al-Bashir is appointed president.

1998 US launches missile attack on a pharmaceutical plant in Khartoum, alleging that it was making materials for chemical weapons.

1998 Devastating famine in Bahar El Gazal; hundreds of thousands die.

1999 President Bashir dissolves the National Assembly and declares a state of emergency following a power struggle with parliamentary speaker Hassan al-Turabi.

1999 Sudan begins to export oil.

2001 *March*: UN's World Food Programme (WFP) seeks funds to feed 3 million facing famine. *April:* SPLA rebels threaten to attack international oil workers brought in to help exploit vast new oil reserves. Government troops accused of trying to drive civilians and rebels from oilfields. *June*: failure of Nairobi peace talks (attended by President al–Bashir and rebel leader John Garang). *November*: US extends unilateral sanctions against Sudan for another year, citing its record on terrorism and rights violations.

2002 *January*: SPLA joins forces with rival militia group, the Sudan People's Defence Force, to pool resources in campaign against the government in Khartoum. Government and SPLA sign landmark Nuba Mountains ceasefire agreement providing for six-month renewable ceasefire in central Nuba Mountains, a key rebel stronghold. *20 July*: after talks in Kenya, the GoS and SPLA sign Machakos Protocol on ending 19-year civil war. Government accepts right of South Sudan to seek self-determination after six-year interim period. Southern rebels accept application of Shariah law in north. *27 July*: President al–Bashir and SPLA leader John Garang meet face-to-face for the first time, through the mediation of Ugandan President Yoweri Museveni. *October*: GoS and SPLA agree to ceasefire for duration of negotiations; despite this, hostilities continue. *November*: negotiations stall over allocation of government and civil service posts, but both sides agree to observe ceasefire.

2003 *February*: rebels in western region of Darfur rise up against the government, claiming the region is being neglected by Khartoum.

2004 *January*: army moves to quell rebel uprising in western region of Darfur; hundreds of thousands of refugees flee to neighbouring Chad. *March*: UN official says pro-government Arab 'Janjaweed' militias are carrying out systematic killings of African villagers in Darfur. *May*: government and southern rebels agree on power-sharing protocols as part of a peace deal to end their long-running conflict. The deal follows earlier breakthroughs on the division of oil and non-oil wealth.

2005 *January*: government and southern rebels sign a peace deal that includes a permanent ceasefire and accords on sharing of wealth and power. *March*: UN Security Council authorizes sanctions against those who violate ceasefire in Darfur and votes to refer those accused of war crimes in Darfur to International Criminal Court. *April*: international donors pledge $4.5 billion in aid to help South Sudan recover from decades of civil war. *9 July*: former southern rebel leader John Garang is sworn in as first Vice-President. A constitution is signed which gives a large degree of autonomy to the South. *1 August*: government announces death of Vice-President and former rebel leader John Garang in helicopter crash; he is succeeded by Salva Kiir. *September*: power-sharing government is formed in Khartoum. *October*: autonomous government is formed in the South, in line with the January 2005 peace deal.

2006 *May*: Khartoum government and a rebel faction in Darfur sign a peace accord. Two rebel groups reject the deal and fighting continues. *August*: Sudan rejects a UN resolution calling for a UN peacekeeping force in Darfur, saying it would compromise Sudanese sovereignty. *November*: hundreds are thought to have died in heavy fighting between northern Sudanese and SPLA forces; fighting centred on the southern town of Malakal. *December*: Sudan agrees in principle to accept the deployment of UN troops in Darfur as part of an expanded peacekeeping force.

References

Deng, L. (1999) *Famine in the Sudan: Causes, Preparedness and Response. A Political, Social and Economic Analysis of the 1998 Bahr el Ghazal famine*, IDS Discussion Paper 369, Institute of Development Studies, Brighton.

Deng, L. (2002) *Confronting Civil War: A Comparative Study of Household Assets Management in Southern Sudan*, IDS Discussion Paper 381, Institute of Development Studies, Brighton.

Deng, L. (2003) 'Confronting Civil War: A Comparative Study of Household Livelihood Strategies in Southern Sudan during the 1990s', PhD thesis, University of Sussex, Brighton.

FAO (Food and Agriculture Organization of the United Nations) (2003) *A Review of Existing Food Security Information Flows in Sudan*, ESAF, Rome, FAO (mimeo).

Harragin, S. (1998) *The Southern Sudan Vulnerability Study*, Save the Children-UK, Nairobi.

Lehtinen, T. (2001) *The European Union's Political and Development Response to Sudan*, ECPDM Discussion Paper 26, European Centre for Development Policy Management, Maastricht.

NSCSE (2004) *Towards a Baseline: Best Estimates of Social Indicators for Southern Sudan*, The New Sudan Centre for Statistics and Evaluation (NSCSE) and UNICEF, Nairobi.

Prendergast, J. and Mozersky, D. (2004) 'Love Thy Neighbor: Regional Intervention in Sudan's Civil War', *Harvard International Review*, 26(1), http://www.crisisgroup.org/home/index.cfm?id=2678&1=1

Russo, L. (2005) *Food Security and Agricultural Rehabilitation with a Medium- to Longer-term Perspective in the Protracted Crisis Context of Sudan* (mimeo).

Save the Children-UK (1998) *An Introduction to Food Economies of Southern Sudan*, SCF-UK, London.

UNDP (2005a) *Human Development Report 2005*, UNDP, New York.

UNDP (2005b) *Sudan: First Interim Millennium Development Goals Report 2004*, UNDP Sudan, Khartoum.

UNICEF (United Nations Children's Fund) (2004) *2003 Consolidated Donor Report Southern Sudan*, UNICEF, New York.

World Bank (2003) *Sudan Stabilization and Reconstruction*, Country Economic Memorandum, World Bank, Washington DC.

WFP (World Food Programme) (2003) *Sudan Annual Needs Assessment 2002/2003*, WFP Sudan, Khartoum.

WFP (2006) *2005 Food Aid Flows*, WFP Interfais, www.wfp.org/interfais/index2.htm [accessed 7 March 2007].

CHAPTER 3

Responding to protracted crises: The principled model of NMPACT in Sudan

Sara Pantuliano

Abstract

This chapter describes the impact of the conflict in Sudan on the Nuba Mountains population and how a parternership between donors, agencies and local stakeholders, based on principles of engagement, resulted in coordinated efforts to address the key determinants of the conflict and food insecurity. Particular attention is paid to the principles of engagement and the 'political humanitarianism' of NMPACT to illustrate how it broke away from the traditional externally driven responses to food insecurity towards an approach that focused on capacity building, sustainable agriculture and market revitalization, alongside conflict transformation and peace-building. Successes, limitations and challenges are distilled to provide lessons for possible replication in other complex emergency contexts.

The Nuba Mountains region: A geo-political overview

The Nuba Mountains are located at the centre of Sudan in the State of South Kordofan and include the six provinces of Kadugli (the state capital), Dilling, Lagawa, Rashad, Abu Jibeha and Talodi. The region covers an area of roughly 80,000 square kilometres (km^2) and prior to the signing of the Comprehensive Peace Agreement (CPA) its population was estimated at between 1.2 and 1.4 million.[1] The main inhabitants of the region are commonly known as the Nuba. This is a highly complex mix of people comprising 50 different groups speaking 50 different languages, who despite this great heterogeneity share a number of fundamental common cultural practices and beliefs, and who widely recognize themselves as Nuba. Culturally and economically the majority of the Nuba are settled farmers, though they share the region with Arab cattle herders, mainly Baggara Hawazma and Shanabla as well as the nomadic Fallata of West African origin (known elsewhere as Fulani). The area has always been recognized as one of the richest and most fertile of Sudan and in the past surplus food production was registered on a fairly regular basis. Unfortunately, the inception of conflict in 1985 and its intensification in

1989 led to a near-total breakdown of the local production system, which has increased the vulnerability of the local population.

The roots of the conflict predate colonial intervention, though the policies promoted by the colonial administration contributed to considerably exacerbate the political and economic marginalization of the people of the Nuba Mountains. Continuing marginalization and discriminatory land policies introduced by various independent governments heightened feelings of frustration and resentment amongst the Nuba people. In the 1970s the abolition of the Native Administration and the introduction of new land laws *de facto* deprived many Nuba of their land in favour of non-Nuba groups and rendered traditional mechanisms of intra- and inter-tribal conflict resolution ineffective. Wealthy northern merchants invested in large mechanized farming schemes on what was previously Nuba land, while local Arab groups invested in small-holders schemes. The mechanized schemes also cut across the transhumance routes of Baggara nomads, who in order to avoid being fined for trespassing frequently re-routed their herds through Nuba farmland. With the absence of a system for settling disputes, armed confrontation started to escalate in the region. The lack of educational opportunities for young people further compounded the feelings of frustration and marginalization amongst Nuba youth at the beginning of the 1980s. Many Nuba became increasingly sympathetic to the plight of the Southerners and decided to support the new civil war when it erupted in 1983 under the leadership of the SPLM/A). The people of the Nuba Mountains entered the civil war in July 1985 led by the late Cdr Yusuf Kuwa, who was an elected member of parliament at the time and was the head of an underground Nuba movement called *Komolo*.

The first incursions of the SPLA in the Nuba Mountains in 1985 sparked a strong reaction from the elected government of Sadiq al-Mahdi, which started to arm Baggara militia as well as Nuba youth forcibly conscripted into the Popular Defence Force (PDF). The militia began a violent and aggressive campaign against Nuba civilians who were indiscriminately accused of supporting the SPLA struggle. In 1988 the government started a policy of systematic elimination of educated Nuba and village leaders, which resulted in an increase in the number of recruits for the SPLA. In 1989 Yusuf Kuwa returned to the Nuba Mountains with a large SPLA force and established a permanent SPLM/A presence in the region, promoted strong political mobilization and reorganized the civil administration in the areas under SPLM/A control (Johnson, 2003). From the late 1980s until the signing of the CPA in 2005 the Nuba Mountains were divided between two administrations, namely the government, which held most of the farmland on the plains as well as the urban centres, and the SPLM/A, which held the crowded hilltops (see Figure 3.1)

Figure 3.1 Map of the Nuba Mountains showing GoS and SPLM areas (2000)

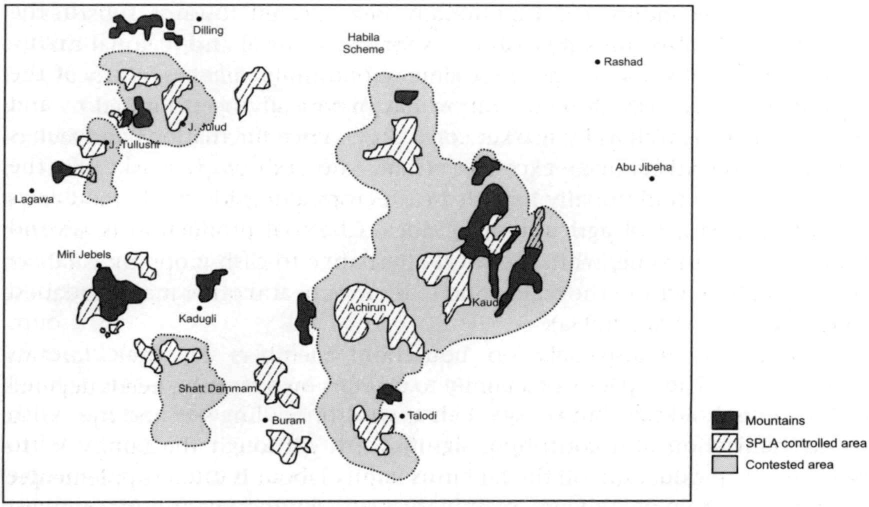

Source: Adapted from African Rights (1995)

Livelihoods systems and food security in the context of crisis in the Nuba Mountains

The farming system

The livelihoods system of the Nuba groups is centred on farming, both in the mountains and on the plains. Four main agricultural systems prevail in the region:

(1) smallholder traditional farming;
(2) mechanized smallholder schemes;
(3) large-scale mechanized farming; and
(4) horticultural production.

The majority of South Kordofan farmers practice traditional smallholder agriculture, which include the following characteristics: small farm areas; subsistence and labour intensive production; no use of machinery, fertilizers, improved varieties or crop protection and primitive production techniques (AACM International, 1993). On the central clay plains and in the eastern and southern parts of the state, a typical Nuba farm is divided into three different fields: house farm (*jubraka*), hillside (near) farm and far farm, according to the literal translation of the vernacular terms used in most Nuba groups (Harragin, 2003a). The *jubraka*, though the smallest, is the most intensively cropped and it is usually the responsibility of women, who also contribute to the other fields. The near farm is often about 2 km from the village, while the far farm can be much further (AACM International, 1993). Crops involve swift maturing varieties of sorghum, maize and beans, as well as groundnuts.

The Nuba economy has traditionally been geared towards subsistence, though people also cultivate cash crops for sale in local and regional market areas. Major Nuba cash crops are sesame, groundnuts, *hibiscus*, cowpeas and watermelon, but cash crop cultivation has historically been limited by lack of technology as well as by market constraints. Price fluctuations and lack of control over markets make excessive reliance on cash crops a risky strategy, so farmers have traditionally included cash crops alongside staple food crops as part of a basket of agricultural produce. Charcoal production is another source of cash income, while a critical alternative to cash cropping is labour migration, both within the region and to Khartoum and other major Sudanese towns (Manger et al., 2003a).

Smallholders mainly rely on household members for their farming requirements. The capacity of a family to meet its own farming needs depends on the household size, but factors such as wealth, holding size and the extent of mechanization also contribute significantly. Although the family is the basic unit of production, on the far farms family labour is often supplemented by assistance from neighbours, mainly through reciprocal communal working parties called *nafirs* (SKRPU, 1980f). The *nafir* is an obligatory institution whereby relatives and neighbours of a family help each other execute labour intensive activities. The importance of non-kin is particularly high when the family moves to a new settlement. In this regard, the institution of *nafir* has played a central role in supporting displaced Nuba families in areas where they had no relatives to count on. *Nafir* participants do not receive any cash payment, but are rewarded in kind. The *nafir* is a distributive mechanism that allows members of the same settlement to express their belonging to a community though reciprocal labour support (Salih, 1984).

The smallholder agricultural system varies slightly for the Arab family farms, which predominantly occupy sandy *qoz* plains in the west of the state. The typical farming Arab household has only two fields: the *jubraka* and a main far field. Both Nuba and Arab smallholders have traditionally kept some animals: goats are the most common among the Nuba (though some also have cattle), while sedentarized Arab groups tend to have sheep and cattle. In some cases their herd sizes can be considerable. Success and failure in the management of animals is a major factor creating differentiation among Nuba households. Successful animal keepers could make agreements with the Baggara nomads on their seasonal migrations to northern Kordofan, thus better exploiting available resources, or some Nuba could even establish themselves as nomads, joining a Baggara camp, though the war has curtailed these strategies (Manger et al, 2003a). The conflict has also severely affected herd ownership patterns and today most farming households are virtually stockless.

Traditional smallholder farming has been complemented by mechanized crop production in parts of the state. Mechanized schemes, which have involved clearing large plots of land, have not been successful and most large schemes have failed. Major reasons for this failure have been the use of follow-on mono-cropping practices, mainly for sorghum and cotton production, with minimal inputs and inappropriate technology (IFAD, 2000).

Constraints to marketing have always been significant in South Kordofan state, particularly given the lack of an adequate road network and market outlets and of appropriate techniques to process or store food. A further factor that prevents smallholders from maximizing the gains of their production is the system of rural credit that dominates in the region, known as *sheil*. The *sheil* system consists of money lenders or merchants who make seasonal advances in cash or in kind to farmers who in turn agree to repay a set amount of produce at a predetermined price (AACM International, 1993). *Sheil* merchants make profits in the region of 40–60 per cent. In addition to the exploitation of the farmers, the *sheil* system is also blamed for hindering agricultural growth in the traditional sector because it gives producers little incentive to increase output as a higher proportion of their gains would go towards the repayment of increased loans (AACM International, 1993). Due to the difficulties farmers face in obtaining formal credit, the *sheil* system remains vital to the seasonal financing of agricultural operations.

The pastoral system

Apart from settled farmers, South Kordofan state is also inhabited by groups of nomadic Arab pastoralists for part of the year. The pastoralists are primarily Baggara Hawazma cattle keepers and Shanabla camel herders and to a lesser extent nomadic Fallata of West African groups (mainly keeping cattle). These groups move over long distances, spending the rainy season in the sandy areas of northern Kordofan and moving southwards into the Nuba Mountains during the dry season, travelling as far as Shilluk land in Upper Nile Province for dry season grazing. These north–south migrations take them through the Nuba Mountains, where they interact with the local Nuba groups.

The cattle herding nomadic groups amount to about 25 per cent of the Nuba Mountains' population, but they own 80 per cent of the livestock (IFAD, 2000), though conflict and drought have significantly affected livestock holdings over the last decade. Nomadic groups spend approximately three months a year on transhumance. In normal rainfall years most nomadic groups end up staying in North Kordofan for about three months before returning to the Nuba Mountains, while in years of poor rainfall they only travel to the northern parts of South Kordofan, where they stay for just six weeks before returning south (IFAD, 2000). Since the signing of the Cease-fire Agreement in 2002 (see below) and even more after the signing of the CPA, some groups have been resuming transhumance along the old routes.

The relations between the nomadic Arab groups and the settled farmers in the Nuba Mountains have been characterized by both peaceful co-existence and confrontation. From a perspective of interacting production systems, settled farming and pastoralism are highly complementary. Until the 1970s in different parts of the Nuba Mountains pastoralists and farmers tried to capitalize on their interaction to maximize the use of available resources. Arab pastoralists were allowed into the Mountains and other farming areas after the harvest was collected and usually stayed there until the first rains. They grazed their livestock on the harvested fields, thus fertilizing them, and helped the villagers transport their grain to the market with their camels. In some cases production and commercial links between farmers and pastoralists developed, with fodder and grazing being exploited after cultivation. Pastoral nomadic populations were therefore fully integrated in the sedentary political economy (Manger et al, 2003a). However, patterns of political marginalization and economic exploitation of Nuba communities have caused relationships in the region to be characterized by conflict rather than complementarity. The last decade and a half of war has further undermined the viability of previous regulatory agreements. Like the settled communities, but for different reasons, pastoralists have also suffered from the establishment of mechanized agriculture schemes (see below), and also tend to be marginalized within wider Sudanese society.

External shocks on food systems and food security

The consequences of the Unregistered Land Act and the expansion of mechanized farming in the region

The land tenure system in the Nuba Mountains has traditionally been based on customary holdings. The system started to undergo important changes with colonial rule.

The British accepted customary rules over land, but the title to land was vested in the government. During the colonial rule the first cotton schemes were introduced in the region both with the aim of growing cheap cotton for the British textile industry and to increase colonial revenues by involving Nuba people in the production of a cash crop that could enable them to pay the poll and crop taxes (Salih, 1984).

After independence the colonial land tenure management system was abolished and tribal leaders were replaced predominantly by northern administrators. Furthermore, the state started to confiscate land to the advantage of wealthy and powerful individuals who started to invest heavily in agricultural schemes in the 1960s. Northern Jallaba traders took control over large portions of Nuba cultivable land, something that created strong resentment amongst the Nuba who started to show signs of revolt during the mid-1960s (Salih, 1995). The Mechanized Farming Corporation (MFC) Act of 1968 established that 60 per cent of land had to be allocated to local people and that no one could have more than one farm, each of which was to be allocated in lots of between 500–1,500 *feddans*.[2] This proviso was ignored however, and some outside landowners ended up with more than 20 farms.

The promulgation of the Unregistered Land Act in 1970, which abolished customary rights of land use, led to deregulation and further seizing of land for agricultural schemes, which cut into prime land of small farmers and nomadic pastoralists. The act did not define the legal status of the current land users and gave the government ample powers of eviction. Compensation for the displaced farmers was discretionary rather than compulsory and often consisted in a choice between inferior land outside the scheme or keeping the existing plot but paying rent for it (Harragin, 2003a). Understandably, very few people were prepared to pay for land they considered theirs.

The Unregistered Land Act provided a legal basis for land acquisition for large-scale mechanized agricultural projects (LTTF, 1986). By 1993 2.5 million *feddans* (over 1 million hectares (ha)) were under mechanized farming and it is estimated that today the figure is in the range of 3–4 million *feddans* (1,260,000–1,680,000 ha), i.e. between 9 and 12 per cent of the total area of pre-CPA South Kordofan (Harragin, 2003a). Considering that all the schemes are on the fertile clay plains, the best soils in the region, which amount to about 21 per cent of the total area of the state, it means that half of the total area of the plains is taken up by the schemes.

The introduction of the Law of Criminal Trespass of 1974 made for even more restrictive rights of access for pastoralists and smallholding farmers to land under schemes. Shortly after the enactment of the Unregistered Land Act, the Native Administration was also abolished with the Local Government Act of 1971, which instituted Executive Councils and subsidiary District Councils and rural, village and nomadic camp councils in all the provinces of the country. These institutions though never became fully functioning and land tenure issues therefore continued to be administered by traditional leaders who no longer had a legal basis to allocate land and solve disputes (LTTF, 1986).

The absence of a regulatory body resulted in sustained land grabbing and intensified disputes between farmers and scheme owners and farmers and pastoralists, transforming traditional tribal animosities into political conflicts, latterly involving the use of modern weapons. The recognition of customary rights was undermined even further by the Civil Transaction Act of 1984, which prohibited the recognition of customary land rights in court. The cumulative effect of the act and measures that had preceded it was to transfer control over land to people connected with those in power and to progressively impoverish rural people (Ajawin and de Waal, 2002; Shazali, 2004).

Economically, the large mechanized schemes yielded considerable profits for many of their owners. In 1979 a calculation of the distribution of incomes on the schemes in the Nuba Mountains between the owners and the workers, i.e. between capital and labour (Manger, 1994), found that 53 per cent went to the owner and 47 per cent to the workers. However, there were only one or two owners, compared with several hundred labourers, so the difference in income distribution was dramatic. The skewed income stream, coupled with the increased vulnerability of the once self-sufficient but now wage-dependent rural poor, further strengthened the already dominant position of northern merchants (Manger et al, 2003a).

Settled farmers were not the only victims of mechanized farming. The mechanized schemes also cut across the transhumance routes of Baggara nomads, who in order to avoid being fined for trespass frequently re-routed their herds through Nuba farmland. In particular, a large number of World Bank supported mechanized farming projects were set up between 1973 and 1993 by the Mechanized Farming Corporation on pastoralists' transhumance routes. This resulted in a lot of conflict between farmers and herders who deviated from traditional routes into Nuba smallholders' land to avoid fines. The most serious problems took place around Habila scheme, which according to IFAD data (2000) today extends across 750,000 *feddans* (315,000 ha).

The abolition of the Native Administration left an institutional vacuum to settle land disputes locally and customarily. Government courts often took the side of the Arab Baggara against the Nuba. Many dispossessed farmers started to seek labour on the schemes or to migrate to northern towns. The lack of educational opportunities for young people further compounded the feelings of frustration and marginalization amongst Nuba youth at the beginning of the 1980s. It is against this backdrop that many Nuba decided to support the new civil war when it erupted in 1983 under the leadership of the SPLM.

The outbreak of conflict in 1985 and its consequences on people's assets and livelihoods

The inception of conflict in 1985 and its escalation in the 1990s led to widespread destruction of traditional sources of livelihoods and massive internal displacement, with few Nuba retaining access to their traditional farming land. This became a key factor in what has become a situation of recurrent food insecurity. Many Nuba ran to the hilltops, where they had no access to the productive clay soils found in the plains. Many areas saw their harvest yields drop approximately ten times (NFSWG, 2001). People were forced to cultivate their main farms on the rocky slopes, in plateaux or next to the mountains, where the soil quality requires heavy labour and where there are restricted areas suitable for cultivation. Livestock rearing was also reduced significantly, since insecurity in the plains made access to pasture land and water points very difficult, especially in the dry season. Looting of cattle also lowered livestock holdings in the areas of the region most affected by the conflict.

The conflict in the Nuba Mountains dramatically changed the pattern and availability of labour opportunities in the region. From the late 1980s until the signing of the CPA the Nuba Mountains remained divided between two administrations, namely the government, which covered most of the farmland on the plains as well as the urban centres, and the SPLM, which covered the hilltops and mountainous terrain. The communities that were most affected were those living in SPLM-controlled areas. Before the war, men would migrate to towns, agricultural schemes and northern markets to look for work. Those who stayed for long periods would send back remittances, but the war cut off this option for those living in SPLM areas, since access to areas under government control was impeded.

Economic isolation was a tactic of the civil war. Access to formal goods markets in SPLM areas was curtailed. Northern traders exploited this isolation by selling goods at high prices in the so-called 'Arab markets' that would take place in the SPLM areas randomly and without a regular pattern whenever northern traders ventured into SPLM areas.

The war also led to a total collapse of social services, including health and education. The number of health facilities and their quality declined markedly over the 1990s, particularly in areas under SPLM control. Table 3.1 shows the differences in availability of health structures between government and SPLM areas.

The conflict also created widespread displacement. In 2003 it was estimated that 636,000 Nuba IDPs lived in government-controlled areas only (IOM/UNDP, 2003). This figure has changed as Nuba IDPs have started returning to South Kordofan state following the signing of the Cease-fire Agreement in 2002 and of the CPA in 2005.

The war was characterized by serious violations of international human rights and humanitarian principles. In many cases civilians were the prime targets of the violations. Raids on villages, farms, settlements and households, expropriation of livestock, abductions, systematic rape, killing and maiming of civilians including the use of landmines, were reported in the region and thoroughly documented by external observers (cf. African Rights, 1995). During the second part of the 1990s the conflict in the Nuba Mountains started to attract widespread international attention both because of the reported human rights violations and because of the blockade on humanitarian assistance imposed by the GoS on the population living in SPLM-controlled areas. In GoS-controlled areas people had access to external assistance including food relief throughout the 1990s.

Table 3.1 Differences in health structures between GoS and SPLM areas

Type of structure	*Ratio population/ structures GOS areas*	*Ratio population/ structures SPLM areas*
Hospitals	128,647	(no hospitals in SPLM areas)
Health Centres	36,972	123,508
Primary Health Care Units	7,980	10,014

Sources: AACM (1993); IFAD (2004b); Office of the UNR/HC (2002a); (2004a).

Changes in food security levels and resilience of food systems

The changes in food security levels over the 1990s in GOS and SPLM areas

The main repercussions of the long years of armed conflict with its consequent displacement of population and destruction of infrastructure were felt in the agriculture and livestock sectors and thus in the food security situation. Production itself was affected by conflict and the previously existing agricultural and livestock support services were eroded to the extent that by the time the war ended they barely existed. Land-use patterns changed during the war, pushing an increasing number of people into distress cultivation on the mountains slopes, especially in SPLM-controlled areas, where a clear relation between the emergence of intensive production systems and the security situation could be observed as a result of the conflict (Manger, 1994).

Land holdings were significantly reduced. In the fertile government-controlled areas of eastern South Kordofan state, where holdings have always been bigger than the rest of the region, also because of the lower population pressure, IFAD (2004a) estimated that the average cultivated area decreased from 34.9 *feddans* (and a maximum of 148) in 1985–1986 to an average area of 30.4 *feddans* (and a maximum of 127) in 2002–2003. In 2000 in the surplus area of El Buram, original villagers were cultivating 3–5 *feddans*, while displaced households only had access to a home garden *(jubraka)* of 0.5–1 *feddan*, leading to food shortages for 60–70 per cent of the total village households (IFAD, 2000).

Crop production also decreased and the ratio between production and consumption sharply changed in GoS areas. Table 3.2 compares data extrapolated from the South Kordofan Rural Planning Unit (SKRPU) for 1980 and from IFAD for 1997–1998.

In the SPLM areas, the amount of land cultivated and the yield per *feddan* decreased for all crops since the war started. Table 3.3 shows the trend for sorghum.

Table 3.2 Per capita staple grain deficit/surplus in GoS areas

	1980	*1997/98*
Average per capita production (kg)	187	103
Average per capita consumption (kg)	139	130
Staple grain balance (kg)	47	–27

Notes: SKRPU data refer to 70 per cent of pre-CPA South Kordofan state. However, the areas not included in the analysis are the eastern provinces, which usually have higher productivity than the state average. IFAD household data assume an average household size of 10.
Source: SKRPU (1980); IFAD, (2000)

Table 3.3 Trends in crop production (sorghum) in SPLM areas

Average household	*Pre-war*	*1999*	*2001*
Land cultivated (*feddans*)	5–7	1–3	0.5–1
Yield of sorghum per feddan (90 kg sacks)	4–5	2–3	1.5–2.5
% total energy requirement available to HH (if all eaten)	190%	27%	11%

Source: Office of the UNR/HC (2002a)

Table 3.3 shows that prior to the conflict the average household was able to secure almost double its food needs from sorghum alone. This allowed a household many options in terms of trade and also meant that there were plenty of labour options available for poor households. By 1999 production had decreased substantially but most households could still meet about one-third of their needs from sorghum consumption, with other needs being met from other food sources. However, by 2001 both yields and the amount of land available had decreased even further, mainly because of insecurity preventing access to land on the plains and because of the resulting increased competition for land on or near the slopes. The decline in yields was undoubtedly due to decreasing soil fertility as 2001 was a very good year in terms of rainfall (Office of the UNR/HC, 2002a). Farmers reported to a UN assessment mission that they no longer left fields fallow or rotated crops and therefore the variety and the quantity of crops grown had decreased. The increased competition over land and the abandonment of the 'shifting cultivation' pattern was a direct result of the displacement of communities from the plains and the insecurity (Office of the UNR/HC, 2002a).

Livestock holdings in the region also decreased significantly as a result of conflict. UN data (Office of the UNR/HC, 2002a) estimated that in SPLM areas holdings had dropped by at least 60– 70 per cent from pre-war levels, with significant losses being observed in GoS areas as well. Most households lost all cattle, both because insecurity in the plains made access to pasture land and water points, essential in the dry season, very difficult, and because of the limited access to livestock drugs in areas where fighting was most intense. Looting of cattle was also a common feature of the conflict. Since large holdings of cattle acted as a target, an increasing number of families chose to keep their herds very small by increasing livestock offtake. This had implications for livestock production but also undermined coping strategies as cattle were traditionally considered a vehicle to preserve wealth as they could be traded for grain in poor harvest years (Office of the UNR/HC, 2002a).

Tables 3.4 and 3.5 summarize the main changes in livestock holdings as a consequence of the conflict in two sample areas in SPLM and GoS-controlled territory.

Changes in relative wealth since the conflict began were also significant both in GoS and even more acutely in SPLM-controlled areas. Wealth ranking

Table 3.4 War-related changes in livestock holdings, Nogorban County (SPLM areas)

Economic status	*No. Cattle*			*No. Shoats*			*% in community*		
	Pre-conflict	*1999*	*2002*	*Pre-conflict*	*1999*	*2002*	*Pre-conflict*	*1999*	*2002*
Rich	50-100	4–10	4	35–50	6–10	5	35–45	6–14	10
Middle	30–50	2–3	2	20–30	4–6	3	18–30	15–20	15
Poor	10–20	0–1	0	10–20	1–3	0	22–30	25–35	75
V. poor	6–10	0	n.a.	6–10	0	n.a.	6–14	36–49	n.a.

Source: Adapted from UNCERO (1999) and CARE (2002)

Table 3.5 War-related changes in livestock holdings, Dilling Province (GoS areas)

Economic status	*No. Cattle*		*No. Shoats*		*% in community*	
	Pre-conflict	*1999*	*Pre-conflict*	*1999*	*Pre-conflict*	*1999*
Rich	200–300	4–8	50–100	10–20	40–52	13–20
Middle	50–100	3–7	30–50	2–7	30–38	30–35
Poor	10–20	0	10–25	0	15–25	45–55

Source: Adapted from UNCERO (1999)

exercises based on crop production, livestock and land holdings show that there was a complete reversal in wealth categories. In Nogorban County perceptions of those falling into the category of being 'better off' stood at 40 per cent prior to the onset of conflict and had collapsed to just 10 per cent in 1999. On the other hand the percentage of very poor increased from 10–42.5 per cent in the same period. In GoS-controlled Dilling the rich were perceived to number 46 per cent pre-conflict and this fell to just 16 per cent by 1999 while the numbers of poor had increased from 20–50 per cent (see Figures 3.2 and 3.3).

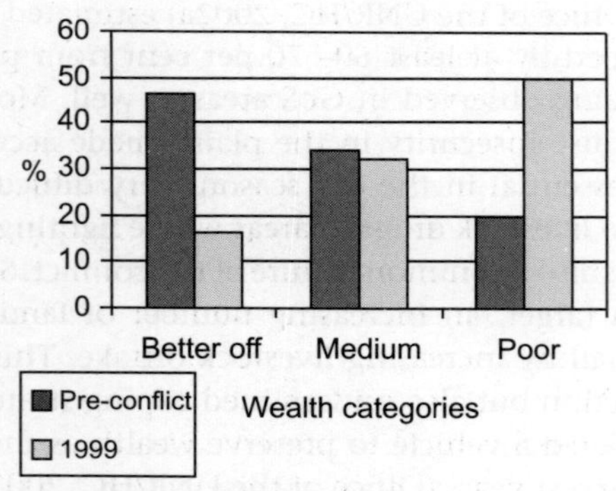

Figure 3.2 Wealth ranking in GoS-controlled areas
Source: Data adapted from UNCERO (1999)

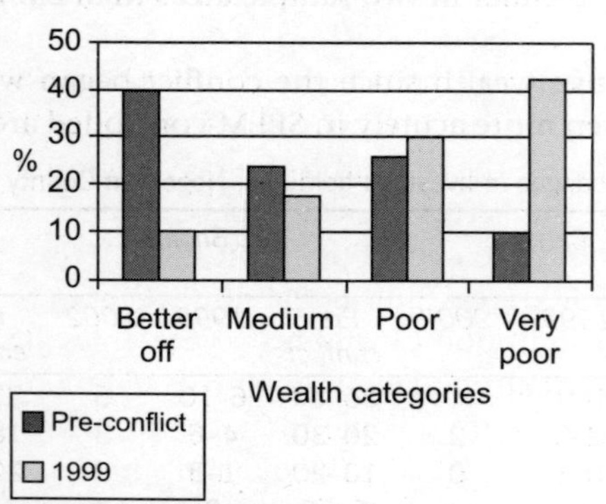

Figure 3.3 Wealth ranking in SPLM-controlled areas
Source: Data adapted from UNCERO (1999)

Indigenous coping mechanisms and the response to external pressure

The lack of economic opportunity and the pressure on farming and livestock holdings caused by the conflict significantly heightened food insecurity for most households in the Nuba Mountains, particularly in the months preceding the main harvest. People became attuned to finding ways of getting over problems associated with a shortfall of the cultivated foodstuffs such as sorghum and maize that are central to the diet. Gathering fruits and wild leaves is extensively practised and during the conflict there was an increase in the importance of wild plants, nuts, fruits, *Acacia* gums, grass grains and tubers as a source of food by the Nuba population.

In 2002 a joint UN/NGO assessment with government and SPLM humanitarian counterparts observed that in the Nuba Mountains the market for gathered foods, fruits, kernels, leaves and roots was thriving (Office of the UNR/HC, 2002f). Some of the products such as *ardeb* (*Tamarindus indica*), *tabaldi* (*Adansonia digitata*), *nabak* (*Ziziphus spinacristi*) and *lalob* (*Balanites aegyptiaca*) were taken by traders to Khartoum and even exported to other countries. Much of the produce would be used for barter, either for imported goods or for grinding sorghum. In 2002 1 *malwa* (3.3 kg) of *gongolese* (*Adansonia digitata*) could be exchanged for 1 pound (lb) of sugar or 0.16 kg of coffee (Office of the UNR/HC, 2002f). Alternatively, the fruits could be sold for cash.

Another important coping mechanism was charcoal making, although this activity was more significant in GoS-held areas where there was more access to woodland on the plains. Prior to the war casual labour opportunities in towns, agricultural schemes and northern markets were an important strategy to cope in times of food stress, particularly during the hunger gap (May–August). However, the isolation of people in the SPLM areas during the conflict restricted the use of local labour markets. Kinship support was also traditionally a key element of the resilience of the Nuba system, understood as the capacity of the system to absorb shocks and adapt to the changes it had been undergoing so as to still essentially retain the same functions, structures, services and knowledge. In SPLM areas during the conflict the chiefs of a community would collect up to 90 kg of cereals from the medium and rich wealth groups after the harvest. The food would be handed over to the Country Administrator who would store it for distribution to the displaced, the returnees, the poor and the very poor during the hunger gap. The contribution of the better off would be voluntary, with each household determining the amount to contribute (UNCERO, 1999).

In the GoS-controlled areas, during the conflict coping strategies in the rainy season included consumption of wild leafy vegetables and various tree leaves and migration for agricultural labour, mainly weeding on mechanized farms. During the dry season many people migrated to towns and to mechanized farms to seek employment, leaving the old and some women behind. Women would also go to Kosti, Abu Jibeha and other towns and work as maids or be engaged in any other available employment. Reductions in the number of meals per day, especially in the hunger gap period was common. Cutting trees for firewood, poles for building and for charcoal making were all widely practised. Kinship support mechanisms were also used, but as the conflict had impoverished all wealth groups, there was little surplus for people to share (UNCERO, 1999). However, food aid from international agencies was available to people in GoS areas throughout the conflict to help them maintain an acceptable food security level.

In the SPLM areas, conversely, the Nuba population received only negligible food aid from a small number of international NGOs that were willing to defy the imposition of the humanitarian blockade imposed by the government on SPLM-controlled areas (see below). Such agencies operated through local institutions, the capacity of which was severely limited to assist the very high number of food insecure people living in SPLM areas. An assessment by the Nuba Food Security Working Group conducted between February and May 2001 estimated that 84,500 people in the region were destitute and lived on a day-to-day basis, with life threatening hunger looming on them during the hunger gap period in the rainy season (NFSWG, 2001). The report, prepared by a number of Nuba officials and international food security experts, played a crucial role in supporting the advocacy campaign that led to the end of the humanitarian blockade in SPLM areas and to the signing of the Cease-fire Agreement in Burgenstock (Switzerland) in January 2002.

The institutional response to livelihoods vulnerability

During the conflict, the food security responses undertaken by local institutions were very limited. In SPLM areas the Civil Authorities developed a welfare strategy that envisaged local purchase of grain and seeds for distribution to the 'most needy' households to supplement other sources of food. The strategy only covered people who were facing the risk of extreme malnutrition that could lead to death or forced migration.

Most of the assistance was brought in and provided by the Nuba Rehabilitation, Relief and Development Organisation (NRRDO), a local NGO set up in 1995 (with strong ties with the SPLM and the Civil Authorities) that enjoyed funding and technical support from a variety of international donors and organizations. NRRDO also undertook limited extension programmes for farmers, but the extent and the quality of both the food relief provision and the agricultural technical support remained extremely limited. NRRDO played a crucial role in discouraging international organizations from delivering excessive quantities of food aid to the area in the wake of the cease-fire and advocated for local purchase of food and seeds as much as possible.

In government areas the local Ministry of Agriculture relied heavily on the provision of food aid by WFP and other international and national organizations to address the needs of IDPs as well as local communities. The quality of the extension services of the ministry had also been progressively deteriorating over the years. International assistance in terms of food aid came to a halt at the end of the 1990s. WFP stopped its operation in the area as a result of the killing of four staff members in June 1998. The agency had been criticized for only assisting populations in GoS areas because of the government ban on delivering aid to SPLM areas. This approach was believed to be encouraging population movement from SPLM areas into GoS areas. The incident sparked much debate amongst international organizations, many of which later decided to withdraw from government-controlled areas until the government agreed to lift its ban on aid delivery to SPLM areas, while others started operating in SPLM areas without permission.

It is important to remark that local authorities on both sides always emphasized that security issues were the primary cause of livelihoods insecurity in the region, which had traditionally been characterized by food surplus in the years before the conflict. In this regard, the cease-fire that was finally brokered in 2002 brought tangible improvements to the quality of life of the people in the Nuba Mountains because increased security allowed people increased access to land and improved trade and access to markets. The concerted action of a number of national and international agencies in supporting livelihoods rehabilitation and strengthening the local food economy in the months following the signing of the cease-fire proved crucial in averting a food security crisis in different areas of the Nuba Mountains.

NMPACT: Beyond conventional humanitarian responses to complex emergencies

The evolution of external interventions in the Nuba Mountains over the 1990s

Following the escalation of the conflict in 1989, the GoS expelled all international NGOs from the Nuba Mountains in 1991 while at the same time intensifying the offensive against the SPLM/A. Soon afterwards, the government imposed a blockade on any relief supplies entering any area under SPLM/A control. The decision was unprecedented in Sudan, since all other areas under SPLM control were covered by the OLS, which distributed relief supplies from its operational base in Kenya.[3] Aid was however allowed in government-controlled areas, particularly in support of the government-controlled 'peace camps' where Nuba people were forcibly relocated *en masse* out of the Nuba Mountains. The massive forced relocation of the Nuba-led human rights organizations to denounce the government policy in the Nuba Mountains as one of 'ethnic cleansing' (African Rights, 1995). The UN estimated that by 1999 there were 72 peace villages in South Kordofan state, with an estimated population of 173,000. UN agencies and a very limited number of NGOs provided assistance to about 105,000 people in 41 peace villages, which were identified as the most vulnerable amongst those affected by displacement (United Nations, 1999).

The humanitarian blockade and the work of humanitarian agencies in government-controlled areas of the Nuba Mountains during the 1990s attracted the criticism of several human rights organizations (Minority Rights International, African Rights, Africa Watch, Human Rights Voice, Amnesty International and Justice Africa amongst others) and sparked much debate within the humanitarian communities in Khartoum and Nairobi. The aid provided by the agencies in GoS areas was seen as instrumental to the government policy of depopulation of the areas under SPLM control and consequently as a factor in the conflict. International agencies like UNICEF, WFP, CARE and UNDP were sharply criticized for their involvement in the peace villages (African Rights, 1995). The blockade to humanitarian assistance in SPLM areas lasted for more than 10 years whilst assistance to government-controlled areas continued unabated throughout the 1990s, though for most agencies interventions were mainly restricted to emergency activities.

All the national organizations operating in the region, with the exception of the Sudan Council of Churches and arguably the Sudanese Red Crescent, were Islamic relief agencies. Indigenous Nuba organizations complained that these agencies were using relief, particularly food aid, to control and Islamize the Nuba. More importantly, it was felt that food was being used as a magnet to force Nuba people out of the SPLM-controlled areas with the promise of food in the peace camps (Rahhal, 2001). But the work of the international agencies received criticism in equal measure, particularly in the case of the two agencies with the biggest programmes in the region, UNICEF and UNDP. Both agencies came under intense criticism by the *OLS Review* (Karim et al, 1996) commissioned in 1996. The review criticized UNICEF for promoting its Child Friendly Village Schemes in 29 villages in South Kordofan, in a context where internal warfare had placed children at great risk. The review wondered to what extent the UN was 'aware of the realities facing the beneficiary populations and the degree to which development initiatives had been explicitly delinked from the political context in which they operated' (Karim et al, 1996).

The review was even more concerned about a programme UNDP was implementing directly with GoS in the Nuba Mountains, the Area Rehabilitation Scheme (ARS) in Kadugli. The *OLS Review* observed that the objectives of the ARS included supporting the local Peace Administration to 'resettle returnees in peace villages and then promote agricultural development to strengthen their attachment to land' (UNDP, 1996, quoted in Karim et al., 1996). The OLS Review Team concluded that given that the Nuba had been dispossessed of their land, the strategy suggested a disturbing ignorance of local realities and that the programme represented a 'de facto accommodation by the UN with disaster-producing policies of the government' (Karim et al, 1996).

Throughout the 1990s the international response in the SPLM-controlled areas was essentially limited to a restricted number of international NGOs funding the main indigenous organization operating in the area, the Nuba Relief, Rehabilitation and Development Organization (NRRDO), which was largely unable to meet the acute needs of the local Nuba population, which became progressively more food insecure.

The increasing use of humanitarian aid as a weapon of war, as with the blockade of assistance to the SPLM areas and the experience of UNICEF and the UNDP ARS in GoS areas, highlighted the need for a more conflict-sensitive approach to programming in the region. Towards the end of the 1990s, the Office of the UN Resident and Humanitarian Co-ordinator (UNR/HC) for Sudan took it upon itself to try and develop a coordinated response for the region, after a period when it promoted intensive efforts to gain access to the SPLM-controlled areas. After years of high-level pressure, which included the involvement of the UN Secretary-General himself in 1998 with an impromptu visit to Khartoum, the UN was finally granted access by the GoS to the SPLM areas to make an assessment in 1999, though a proper humanitarian intervention did not begin until 2002.

The findings of the 1999 inter-agency mission, which visited both SPLM- and GoS-controlled areas, emphasized that assistance to the Nuba Mountains population would be best provided through a comprehensive, multi-sectoral, multi-agency rehabilitation programme addressing both SPLM and GoS-controlled areas, implemented outside the OLS structure, both for reasons of expediency, given the government's strong opposition to extending OLS to the Nuba Mountains, and to identify a response that was more appropriate to the Nuba Mountains context. The political and security situation in the Nuba Mountains prevailing at the end of the 1990s was such that a humanitarian response was required that took into account the difficulty of operating in a complex political environment where humanitarian aid was being used as a weapon in the conflict. It had become apparent to many of the actors involved that only a concerted effort based on policy dialogue with the parties to the conflict and with key external players could have unblocked the impasse around the provision of humanitarian assistance to the region.

NMPACT: Key features

Following the 1999 assessment, a consultative process with a wide range of international NGOs and UN agencies with interest in the Nuba Mountains was started in January 2000 under the leadership of the Office of the UNR/HC, to design the Nuba Mountains Programme. The process was highly inclusive and several meetings were held with all partners involved in the Nuba Mountains, Khartoum and Nairobi with the aim of building a common platform amongst actors, both national and international, who had long been working on the opposite sides of the political divide. After a year-long consultation process with programme partners, a joint programme document was endorsed in May 2001, where emphasis was placed on the development of a set of principles of engagement to be adhered to by all agencies. The implementation of the Nuba Mountains Programme was however hindered by the stalemate over the issue of access to SPLM-controlled areas, which continued to be denied by the government despite repeated promises to the highest levels in the UN. The programme agencies therefore decided to focus their efforts on advocacy

Table 3.6 NMPACT Principles of Engagement

Principles of engagement
All interventions part of a single, integrated, conflict transformation programme
Develop an enabling environment for Nuba-led longer-term peace process
Use 'least harm' approach – avoid endangering opportunities for longer-term peace building
Ensure that interventions strengthen self-reliance, local capacities and opportunities for socio-economic and cultural interdependence
Ensure protection of human rights and sources of livelihoods
Be flexible: responsiveness to changing conditions
Obtain unimpeded, secure access to all areas in Nuba

directed at Western diplomats to facilitate unblocking the humanitarian impasse in the Nuba Mountains, particularly in light of the fact that a food security crisis was maturing in SPLM-controlled areas.

The advocacy action was a major factor in catalysing senior diplomatic interest that in January 2002 resulted in the brokering of the Cease-fire Agreement. The NMP consultation process was extended to all the agencies with an interest to operate in the Nuba Mountains region and benefited from the strong involvement of Nuba partners from various civil society organizations. The new initiative came to be known as the Nuba Mountains Programme Advancing Conflict Transformation (NMPACT). NMPACT was designed as a phased, multi-agency, cross-line programme aimed at enabling all stakeholders to contribute to a Nuba-led response to address the short and long term needs of the people of the Nuba Mountains. Its overall strategic goal was: 'To enhance the Nuba people's capacity for self reliance within a sustained process of conflict transformation guided by the aspirations, priorities and analyses of the Nuba people themselves.' As specified in the strategic goal, the primary target groups of the programme were the Nuba communities, especially in areas of greatest needs. Given the focus of the conflict and the historical marginalization of Nuba communities in the region, the overall goal was formulated to give special emphasis to the Nuba people's role in guiding the programme (Office of the UNR/HC, 2002b).

The programme constituted a major breakthrough in that it became the first and only programme to be subscribed to by both the GoS and the SPLM while the conflict was still in an active state. The GoS Humanitarian Aid Commission (HAC) and the SPLA/M Sudan Relief and Rehabilitation Association – later renamed the Sudan Relief and Rehabilitation Commission (SRRC) – were included as equal partners in the NMPACT Coordination Structure together with an international programme coordinator. Such an institutional set up was unprecedented in Sudan's humanitarian context. The full involvement of HAC and SRRC in the coordination structure gave them a strong sense of buy-in into the programme, towards which they consistently showed strong commitment and interest in facilitating its speedy implementation. The Coordination Structure was also made up of field coordinators in both the

GoS and the SPLM areas who worked equally closely with their respective HAC and SRRC counterparts. Many regard the involvement of the warring parties in a single programme and the cross-line focus of the initiative as the most significant achievements of NMPACT (Office of the UNR/HC, 2003).

The extensive consultation process that had accompanied its development produced a large amount of consensus. By the end of 2003, nine UN agencies, 16 international NGOs and 24 national NGOs had endorsed the programme. Seven of the partners took an active role in becoming sectoral focal points for the NMPACT programming sectors, which included: Agriculture and Food Economy, Education, Health and Nutrition, Water and Environmental Sanitation, Livelihoods Rehabilitation and Peace Building. The Coordination Structure was able to benefit from the technical support of two advisers assigned by USAID who were specialized in agriculture and food economy and in land and natural resource issues.

The extensive consultation process undertaken to design NMPACT also actively involved a high number of donors in drawing up the programme framework. This approach proved to be extremely useful in gaining the buy-in of the donors from the start and to ensure that key elements of the programme were funded as implementation began. Although funding gaps remained important for some agencies, particularly within the UN family, the level of funds allocated to NMPACT partners was highly significant, totalling in excess of $18 million in its first year of implementation (Office of the UNR/HC, 2002c).

The highly participatory approach adopted by NMPACT was reflected also in the design of a policy-making structure that would support the Coordination Structure in orienting collective decision making. A mechanisms was created that allowed all implementing partners to meet systematically at a neutral location in the Nuba Mountains in what was called the 'NMPACT Partners' Forum' (see below).

The OLS: Lessons learned and its implications for NMPACT

OLS had been operating for more than a decade with two separate structures in GoS and SPLM-controlled areas and there was a high level of mistrust between the international organizations working on the two sides of the political divide, let alone the parties at war.

The task of lowering the level of suspicion between the warring parties and the international partners working on the two sides of the political divide proved to be a major obstacle and required a considerable investment in staff time on the part of the Office of the UNR/HC, including the UNR/HC himself, to ensure that the consultation process was genuinely participatory and that consensus around the initiative was maximized amongst the potential partners.

NMPACT was able to capitalize on the lessons learned from OLS and to build on the criticism that this had received from various quarters (Karim et al, 1996; African Rights, 1997). OLS was developed as an access mechanism to allow a rapid response to a critical humanitarian crisis in the South at the end of the 1980s, and it then gradually became an umbrella for coordinated programming as well, while NMPACT set out from the start as a joint coordinated programming framework. The main lesson learned from the OLS was obviously to transcend the North/South divide and to establish one single, coordinated cross-line initiative. NMPACT therefore constituted a departure from the mode of coordination offered by the OLS in that it was the first substantial attempt to bridge the long-established division between agencies based out of Khartoum and Nairobi. The change in approach enabled the programme to attract the involvement of a high number of NGOs, many of which had refused to join OLS and which were not part of its consortium, with only two NGOs operating in the Nuba Mountains and the International Fund for Agricultural Development (IFAD) remaining outside the NMPACT framework. These three agencies however liaised closely with the NMPACT partners, attended the fora and provided the partners with logistical support when needed.

The Coordination Structure designed by NMPACT also departed from the OLS model in the way it involved the official government and SPLM counterparts. Relations between OLS and its humanitarian and political counterparts had often been strained, both in the North and in the South, with the government and the SPLM frequently being obstructive and displaying dissatisfaction for the operation (Karim et al, 1996). The NMPACT strategy of fully involving HAC and SRRC together in the coordination and implementation of the programme proved to be successful. By working together around a common platform HAC and SRRC neutralized each other's more extreme positions and engaged with the international partners in a constructive manner. Bringing together key actors working on the two sides of the political divide into the programme helped to create a new environment of trust and collaboration that spilled over to other areas of assistance in Sudan.

Another distinctive difference between NMPACT and OLS was that coordination was based upon a set of principles of engagement (see below). These principles were developed by the NMPACT partners and Nuba representatives and provided a solid programmatic framework.

The principles of engagement

Much of the uniqueness and effectiveness of NMPACT derived from the principles of engagement. These provided the partners with an overall framework to buy into and gave the joint response a strong conceptual rootedness. The development of the principles stemmed from the common analysis of the partners of the limitations of traditional approaches to complex emergencies founded on the humanitarian principles of neutrality and impartiality. The experience of the external interventions in the Nuba Mountains over the 1990s had created a shared understanding between the

NMPACT stakeholders of the political functions of aid in conflict situations (Macrae and Leader, 2000). This common understanding led to the articulation of the 'principles of engagement', the underlying theme of which was to integrate the aid framework within a political framework to operate in a conflict context. The NMPACT principles of engagement can be summarized as (Office of UNR/HC, 2002b):

- All interventions to be part of a single, integrated, conflict transformation programme;
- Develop an enabling environment for a Nuba-led longer term peace process;
- Use 'least harm' approach – avoid endangering opportunities for longer-term peace building;
- Ensure that interventions strengthen self-reliance, local capacities and opportunities for socio-economic and cultural interdependence;
- Ensure protection of human rights and sources of livelihoods;
- Be flexible and responsive to changing conditions; and
- Obtain unimpeded, secure access to all areas in Nuba.

Though it has been difficult to assess the level of success of the Coordination Structure in ensuring partners' adherence to all the principles, these are regarded by all involved as providing an extremely valuable programming tool. The principles focused on sustainability of programmes, national ownership, equitability of interventions across the political divide, transforming conflict and 'doing least harm', as the 'do no harm' approach (Anderson, 1999) was renamed by the NMPACT partners. The principles of engagement represented an innovative instrument of aid coordination in the context of assistance to Sudan, especially in areas affected by conflict.

The NMPACT internal review of 2003 emphasized that, thanks to the principles of engagement, such as the focus on capacity building, NMPACT had been effective in generating a strong sustainability focus that cut across the work of the partners and that had resulted in the implementation of programmes that were directed more towards training and capacity building than to the delivery of external inputs (Office of the UNR/HC, 2003). This trait is particularly significant given the fact that agencies were operating in an environment where the cease-fire had not yet matured into a peace agreement and represented an important departure from the model of assistance used in other areas of conflict in Sudan.

One of the most fundamental principles of engagement was that of equitability. The principle advocates for the use of measurable and fair standards to ensure that partners' interventions respond to local needs and capacities without re-enforcing the underlying causes of conflict. In order to provide the partners with an objective basis to apply the principle, the First NMPACT Partners' Forum recommended that a region-wide cross-line survey be undertaken in order to provide the partners with the necessary data and information to prioritize areas of intervention and target the population in an equitable manner. The survey, carried out by the partners and their counterparts in November 2002, aimed to analyse strategies and goals of the Nuba people and the barriers they faced, especially with regard to return, resettlement and recovery, in order to understand the socio-economic and political contexts of the possible interventions of the NMPACT partners, and to collect sufficient information to compare livelihoods and geographic differences in people's quality of life in order to support the principle of promoting equitable and fair interventions.

The data collected during the survey showed that there was a profound gap in terms of access to facilities, with communities in SPLM areas being distinctly disadvantaged compared to those in GoS areas. However, the survey report emphasized that the key element for the NMPACT partners was not the provision of services, as most of the people interviewed were still affected by the main consequence of the crisis in the Nuba Mountains: displacement. The survey team argued that for the process of rehabilitation to be sustainable, provision of services and other type of assistance had to be linked to people's return to their land, as this was the only strategy that would have allowed people to have access to a sustainable livelihoods resource base and to take advantage of existing economic opportunities. The results of the survey were presented to the Second Partners' Forum, where the partners decided to collectively embark on a series of studies on land tenure to inform partners' efforts to support the return of IDPs (Office of the UNR/HC, 2002e).

The principles of engagement indirectly became an important instrument to formulate policies, as the information collected to underpin the implementation of the principles had an inevitable impact on the policy-making processes within the programme, resulting in the prioritization of the issues of displacement and land tenure. Other principles, such as that of the protection of sources of livelihoods, drove the Coordination Structure jointly with some NMPACT partners to formulate clear environmental guidelines (including specific procedures for dam construction) to be adopted by the NMPACT partners (White, 2003).

Lastly, the principle of supporting national ownership made NMPACT unique in its involvement of government and SPLM counterparts in the coordination of the programme, thereby conferring ownership of the process to the national authorities. Local ownership was also reinforced through the participation of a large number of national representatives in the partners' fora, where key programming decisions were discussed and agreed upon. The fora, as well as other cross-line meetings, were held in a neutral location in the Nuba Mountains established with the consensus of both warring parties. The fact that NMPACT brought the GoS and the SPLM together on Sudanese soil several times in a neutral environment has been seen by many programme stakeholders as a substantial contribution to the conflict transformation process in the region, which remains the ultimate goal of NMPACT.

'Political humanitarianism' and collective advocacy

The process of programme design for NMPACT went hand in hand with a strong and coordinated advocacy action directed at Western diplomats to facilitate the unblocking of the humanitarian impasse in the Nuba Mountains. This had particular significance in light of the fact that a food security crisis was evolving in SPLM-controlled areas. Such action culminated in the collective decision between 2000 and 2001 of most of the agencies operating in GoS-controlled areas either to suspend their operations in the North or to initiate activities in SPLM-controlled areas where access was denied by the GoS. This move was aimed at applying pressure on government officials to open up access to SPLM-controlled areas, where needs were known to be great and increasingly acute. The decision to withdraw from GoS areas was difficult to take, as this *de facto* meant depriving more needy people of external assistance, but the common analysis of the partners was that aid was being used to lure away people from SPLM areas into GoS areas, thus contributing to exacerbate the conflict in the region. For this reason, it was felt that temporary withdrawal from government-controlled areas was the most ethical short-term choice.

The partners were aware that the mounting crisis in SPLM territory required a political solution and that they needed to attract more international attention to the situation in the Nuba Mountains to resolve the access issue. The UNR/HC at the time therefore used his offices to increase advocacy with western diplomats on behalf of all the partners. This action was a major factor in catalysing senior diplomatic interest that in January 2002 resulted in the brokering of the Cease-fire Agreement. The accord was aided by the offices of US Senator John Danforth, who had been appointed US Envoy for Peace in Sudan by President Bush on 6 September 2001. The signing of the agreement presented those involved in the Nuba Mountains with a major opportunity. The NMPACT programme finally had a chance to become operational. In its final design it became closely linked to the implementation of the Cease-fire Agreement and stipulated close cooperation with the Joint Military Commission/Joint Monitoring Mission (JMC/JMM), the international force mandated to monitor the cease-fire as well as the military and policing roles of the parties in the region. Once again, this represented a novel development in the context of Sudan in that a humanitarian intervention was expressly linked to a political initiative.

The vigorous interaction with key political and military actors involved in the Nuba Mountains was an important constant of the NMPACT approach. From its very inception NMPACT was actively engaged with the JMC/JMM and there was regular and structured interaction between NMPACT and the Friends of Nuba Mountains, a group made up of senior diplomats working in the Sudan, which provided political leadership for the JMC/JMM. The actors concerned, particularly the JMC/JMM, were not always entirely amenable to the concerns raised by NMPACT. However, a deliberate commitment to active, constructive engagement cemented relations and over time proved crucial in ensuring that a number of important issues, which are beyond the remit of humanitarian organizations but that impacted on the response, were addressed in a timely and adequate manner. These included land tenure issues, conflict between nomadic and farming groups and the harassment by the authorities of civilians returning to farms (Office of the UNR/HC, 2003).

The multiple forms of advocacy and engagement with a range of national and international political bodies promoted by the UN agencies and the partner NGOs since 1999 allowed NMPACT unprecedented links, on the part of a humanitarian operation, to the political sphere, an approach that was defined as 'political humanitarianism' (Pantuliano, 2003). Some of the partners argued that particularly in the early period of the Cease-fire Agreement, NMPACT was a key factor underpinning the first extension of the cease-fire since it was seen as an important element of the peace dividend (Office of the UNR/HC, 2003). Later on, NMPACT's research work on land tenure issues (Alden Wily, 2005) was used to inform the special negotiations on the contested areas that took place in Kenya from January 2003 to January 2005 within the context of the wider Sudan peace process. In addition, the studies provided the basis for developing the Terms of Reference of the Nuba Mountains Land Commission envisaged by the Two Areas Protocol regulating peace in the Nuba Mountains and Southern Blue Nile, agreed in Naivasha, Kenya, in May 2004 and endorsed in the implementation modalities of the CPA signed in January 2005.

Food security and land tenure

The vigorous advocacy action that had been promoted as a result of the collective adherence of NMPACT partners to the principle of 'do no harm' (Anderson, 1999) to obtain a cease-fire agreement in the region had largely been prompted by the need to avert a severe food security crisis looming over the SPLM-controlled areas of the Nuba Mountains. These areas had not received international assistance since 1989 and there was therefore a danger of destabilizing the local economy and creating a dependency syndrome through the provision of food aid, as had happened in many parts of southern Sudan. A new approach was designed within NMPACT where food delivery was coupled with programme interventions strongly focused on supporting local capacity and enhancing sustainability through strengthening the local food economy.

The NMPACT food security approach prioritized capacity building over the delivery of external inputs (food aid and infrastructure) and removal of the constraints to food security (insecurity, barriers to access to land, market constraints, amongst others) from the onset of the intervention. Delivery of food aid and seeds and tools also took place in the Nuba Mountains the context of NMPACT to support more vulnerable communities, but these interventions were coupled by joint efforts to root the partners' response into a deeper understanding of the causes behind food insecurity in the region.

The population of the Nuba Mountains was subdivided by the NMPACT partners according to the livelihoods activities in which people were engaged, i.e. rural farmers (in GoS and SPLM areas), pastoralists, urban dwellers and IDP camp occupants, the latter three categories only found in GoS areas (Office of the UNR/HC, 2002d). The rural farmers were later divided between poor, average and better off depending on their holdings (Office of the UNR/HC, 2002f). The principle of equitable assistance, which was one of the fundamental principles of engagement of NMPACT, required that assistance be provided in an equitable manner on the basis of need. This meant that the partners had to prioritize camp occupants and farmers, who had been identified as the most vulnerable groups, in removing barriers and recovering assets to rebuild their livelihoods security. The findings of the cross-line survey in late 2002 highlighted the need to address the issue of displacement within the Nuba Mountains as a priority, particularly for people confined to IDP camps, in order to facilitate people's return to their homeland and their access to a sustainable resource base.

The partners' fora and the cross-line survey also showed the need for the partners to place a special focus on land tenure issues, which were perceived to be one of the greatest constraints to food security in a region that had been considered largely food secure in the past. Several studies were carried out (Manger et al, 2003a; Manger et al, 2003b; Harragin, 2003a), including a three months survey that covered all parts of the Nuba Mountains region (Harragin, 2003b). The survey analysed and recorded traditional land ownership, existing land titles and illegal land alienation to non-Nuba owners. This work was undertaken in order to underpin advocacy action to ensure that IDPs could reclaim land grabbed in the past and return to their farms in contested areas of the Nuba Mountains or receive compensation. It is important to emphasize that the research work on land tenure was carried out while the conflict was still active, albeit under conditions of cease-fire.

Advocacy action was promoted by the NMPACT partners to ensure that local purchase of food from within the Nuba Mountains was maximized and that food aid was limited to areas of extreme need where cultivation had not been possible. The advocacy action brought limited results during the first two years of operation of NMPACT but was successful in ensuring an adequate targeting of communities and more strategic use of food aid.

Analysis of NMPACT food security using the FAO twin-track approach

FAO has developed an analytical framework that aims to assess the health of a food system in crisis. This is an attempt to help those responding to food emergencies to consider their interventions in terms of the resilience of the system to withstand shocks in the longer term and in so doing think well beyond the immediate and temporary efficacy of emergency responses to immediate and life saving needs. Elements of such resilience include 'strengthening diversity; rebuilding local institutions and traditional support mechanisms; reinforcing local knowledge and building on farmers' capacity to adapt and reorganise' (Pingali et al, 2005).

The framework is organized in terms of two 'tracks' that are considered mutually reinforcing. Direct and immediate access to food is the first and is what is considered essential in the immediate term and important for medium-term planning. Rural development and product enhancement is the second track and consists of elements that its creators consider are essential for stability and predictability.

The twin-track framework, which is premised on the assumption that food emergencies are social and political constructions, is consistent with the thinking that underpinned NMPACT. An analysis of the NMPACT interventions using the twin-track framework is therefore useful in determining the extent to which the programme lived up to its objectives. Figure 3.4 shows how the food security related interventions of the NMPACT partners evolved over the period 2002 to 2004 and are compared with the state of interventions of the same organizations prior to the establishment of NMPACT in 2002.

The summary of the 188 interventions covering the work of 14 NMPACT partners involved in the agriculture and food economy sector, shows a number of clear trends.[4] Key to these is that since the inception of the programme the balance of interventions increasingly falls into the category of 'rural development and productivity enhancement' in both GoS and SPLM-controlled areas as opposed to those that are described as belonging to 'direct and immediate access to food'. This is significant given that until the beginning of 2002, major parts of the Nuba Mountains were under an effective aid embargo and the region was in the midst of conflict. In other

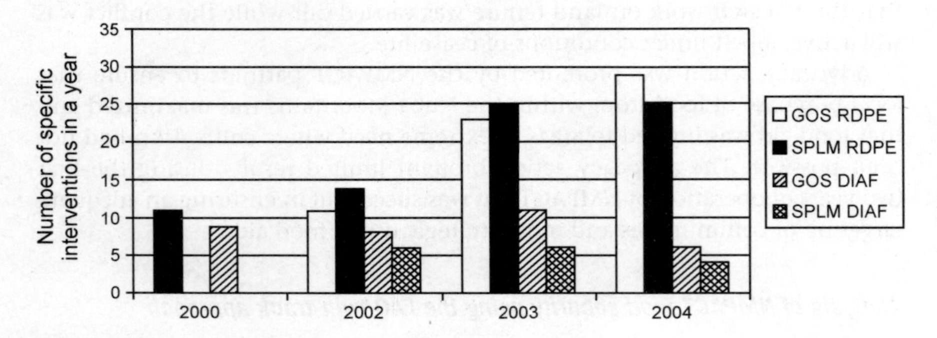

Figure 3.4 Analysis of NMPACT using twin-track approach
Note: RDPE = rural development and productivity enhancement; DIAF = direct and immediate access to food.
Source: Information derived from a series of NMPACT documents, chiefly the information tables produced between 2002 and 2004, and from a stocktaking exercise detailing agencies' activities in South Kordofan State, which was prepared during the development of the Nuba Mountains Programme, NMPACT's precursor. The information tables can be found in Pantuliano (2005).

circumstances the trends would be quite different, but here it would appear that NMPACT, with its emphasis on local capacity building, sustainability and protection of livelihoods, delivered in its own terms, and in line with the twin-track approach successfully facilitated a collective response that buttressed the stability of the food system. The direct and immediate access to food element remained fairly constant in terms of the numbers of interventions, though showed signs of tailing off in 2004. The modest nature of this element of the response in a crisis of this nature and magnitude is likely to be unusual (and for example is in direct contrast with what happened under OLS) given the tendency for agencies to solicit as well as receive encouragement to provide food and other short-term emergency provisions such as seeds and tools, which are part of this framework.

From the NMPACT information tables it is difficult to assess the full extent of the impact on the ground of the collective NMPACT partners' intervention in support of the recovery of local food systems, since a full impact assessment is yet to be undertaken. However, at the peers review workshop organized by FAO in Nairobi in January 2005 during the preparation of this study, representatives from Nuba communities and international food security experts working in the Nuba Mountains emphasized that significant changes have taken place in the region since NMPACT became operational. Some of the examples quoted included the increase in the number of markets throughout the region, the levelling of prices between markets in GoS and in SPLM areas (in 2001 market prices for non-locally produced goods in SPLM areas were at least double the prices in GoS areas), the increased diversity and availability of goods in SPLM markets, the opening of cattle markets and increased market access for farmers and livestock keepers. Participants also mentioned improved access to key services such as water.

These preliminary observations, which obviously will need to be corroborated by in-depth research and analysis, seem to suggest that NMPACT's approach to food security had an important role in strengthening people's own strategies to enhancing resilience and lowering the dependency on external food aid, as the decrease in the number of agencies involved in emergency delivery of aid and seeds seems to demonstrate. It was commented at the peers review meeting in January 2005 that NMPACT's innovative food security approach was made possible because it was part of a wider institutional context where local counterparts were genuinely committed to promoting more long-sighted responses and not to manipulating external emergency assistance for political purposes. Undoubtedly, NRRDO's role in discouraging international organizations from delivering excessive quantities of aid to the Nuba Mountains in the wake of the cease-fire and its advocacy in favour of local purchase of food and seed played a crucial role in shaping the design of NMPACT and its food security strategy.

The interface between local institutions and external stakeholders

Since its formation, NMPACT strove to promote Nuba leadership in the implementation effort and to confer ownership of the implementation process to the national authorities. The SPLM-controlled areas of the Nuba Mountains had developed a remarkable and unique experiment in grassroots democracy that was unparalleled in the rest of the country, be it in government or SPLM administered areas. This was largely thanks to the vision of the late Yusuf Kuwa Mekki, the first SPLM governor of the Nuba Mountains, who endeavoured to initiate a democratic political process in the areas under his control. The centrepiece of such process is the South Kordofan Advisory Council, a Nuba parliament that has been meeting yearly to decide on the most important matters of policy facing the Nuba (cf. Flint, 2001). The council, established in 1992, was the supreme legislative body in the SPLM areas of the Nuba Mountains and had the authority to overrule the executive (the governor). A functioning judiciary was also in place in the SPLM areas. This form of collective, democratic decision making was a remarkable achievement in the context of Sudan, especially in an area that was at war for nearly two decades, and the NMPACT partners were committed to ensuring that the programme would not undermine emerging Nuba institutions.

The strong involvement of HAC and SRRC in the Coordination Structure provided the partners with a channel to address issues with official counterparts both at the field and central (Khartoum/Nairobi) levels, thereby facilitating prompt resolution of problems when they arose. Although the programme did well to involve government and SPLM counterparts in the coordination of the programme, the Coordination Structure and the partners were not equally successful in extending this ownership to the Nuba NGOs and the community on the ground during the first phase of the programme. The aim of promoting genuine Nuba leadership within the response as a whole therefore remained elusive. The lack of local Nuba control over the interventions that were being designed and carried out was a flaw that came to the surface as the programme was rolled out. While many partners focused their efforts on capacity building of local communities, very little was done to support the emergence of genuine Nuba leadership, as envisaged by the NMPACT document. This limited the capacity of the local communities to steer the rehabilitation and development process and the ability of the partners to focus their response in line with a genuinely Nuba analysis, set of aspirations and priorities. The imbalance of power was skewed in favour of international humanitarian representatives when it came to setting agendas and priorities for the interventions in the region, including food security responses. However, many of the NMPACT partners recognized that it was incumbent upon them to remedy this situation in order to be true to the philosophy and mandate of the programme (Office of the UNR/HC, 2003).

More efforts were undertaken at a later stage to involve the Nuba at the grassroots level in all phases of the programme cycle. A NMPACT Monitoring and Evaluation Unit made up of staff from the Nuba Mountains was set up with the support of the World Bank, which trained Nuba Mountains communities in participatory planning, monitoring and evaluation techniques (World Bank, 2004). The underlying idea was that trained communities will be empowered to set priorities for rehabilitation and development interventions in their areas, monitor implementation of programmes and projects and review the performance of external agencies vis-à-vis the principles of engagement.

Information flows and links with the NMPACT response and policy framework

The success of NMPACT in its early days was due in large part to the fact that the programme had a dedicated coordination structure at both the local and the central levels that facilitated the flow of information between the partners. In the 2003 internal review many of the NMPACT partners observed that the NMPACT framework and the Coordination Structure had been instrumental in helping them define, prioritize and coordinate activities.

Within the programme, information was mainly shared though circulation of written material via the Coordination Structure as well as through personal interaction. Regular reports and in-depth studies were circulated to the partners by the Coordination Structure, which would also circulate partners' document to the whole range of partners. Furthermore, a detailed 'NMPACT partners' information table' was regularly prepared and shared with all programme stakeholders, including donors.

Attempts to create a database accessible to all partners and stakeholders were also made following the conclusion of the Baseline Data Collection Survey in November 2002, during which team members were able to gather a high amount of data for each of NMPACT's technical sectors. However, the establishment of the database was hampered by the turn over of personnel in the coordination of the programme.

Regular monthly meetings of the partners were scheduled in Khartoum and in Nairobi as well as at the field level, both in Kadugli (GoS headquarters) and Kaoda (SPLM headquarters), with the main aim of sharing information and reviewing progress towards the implementation of the principles.

The most important avenue for information sharing was obviously the partners' fora, where all stakeholders both at the capitals and the field levels were gathered together to review progress, share information and discuss policy issues. The fora provided an invaluable opportunity for national and international partners operating at the local level to meet in the same place with managers, donors and policy makers stationed in Khartoum and in Nairobi and take joint decisions on key aspects of the programme. This meant that Nuba people from local CBOs and NGOs had a chance to actively influence and direct the NMPACT policy agenda and orient the priorities of the programme. At the fora the partners would collectively review the implementation of the planned activities, share information and discuss the collective research agenda to inform policies aimed at strengthening partners' interventions.

For example, at the First Partners' Forum in July 2002 the decision was made that more analysis and investigation was needed to identify disparities and different levels of needs in the region and prioritize interventions on actual needs (Office of the UNR/HC, 2002d). This led the partners to plan and carry out the region-wide cross-line survey that provided the basis for the second phase of NMPACT, which was focused on rehabilitation. At the Second Partners' Forum in December 2002, a collective decision was made that more research was needed into the issue of land tenure, also to underpin the results of the cross-line survey, which had identified displacement and return as the most critical issues to be addressed by the NMPACT partners (Office of the UNR/HC, 2002e).

Limitations in delivering the model and new challenges

Institutional failings and their effects on implementation

The central role of the Coordination Structure in the success of NMPACT was further brought to light by a year-long staffing gap in 2003, both at the central and field levels, which was largely the result of bureaucratic and administrative delays of both UNDP and UN-OCHA (Office of the UNR/HC, 2003). This gap left the programme without leadership and support and especially affected the partners' focus on the principles of engagement and the interaction between the counterparts. The absence of field coordinators on the ground led counterparts and partners to complain that insufficient attention was being paid to peripheral areas of the Nuba Mountains region, with the consequence that the 'doing least harm' principle was neglected (Office of the UNR/HC, 2003). The resultant lack of information on needs and disparities undermined the development of the intended focus on equitable responses across the region, particularly along political lines and for the different livelihoods groups. Furthermore, the prolonged lack of field coordinators weakened the capacity building process of HAC and SRRC, frustrating their efforts to play their coordinating role effectively, as well as undermining attempts to root the response more deeply amongst a diverse set of local actors.

Crucially, collective decision making, which had so marked the evolution of NMPACT, was restricted by a change of leadership within the UN system, which put strong emphasis on the internal coherence of UN activities and structures. In an attempt to restructure the UN operation throughout Sudan, unilateral decisions about the NMPACT programme were made that did not fully involve either the counterparts or the partners. This had negative effects on the trust building that had been forged in the preceding years. In particular, the official counterparts were disappointed with this turn of events and over time relations gradually deteriorated. Both parties disliked the change of approach and the SPLM in particular felt that certain decisions had considerably affected their interests.

The absence of a fully functioning Coordination Structure was felt particularly in relation to the monitoring of the principles, especially that of equitability. It is interesting to note that the JMC/JMM commented that in general terms they perceived the NGOs as having better incorporated the NMPACT principles into their operation than the UN agencies, whose adherence to the principles diminished once the Coordination Structure became less operational (Office of the UNR/HC, 2003).

The changes that arose around the implementation of NMPACT reflect weaknesses within the UN coordination system as a whole. NMPACT was born out of the vision of an array of national and international actors and many within the UN system provided it with leadership. Despite the presence of a wide number of influential backers, ranging from donors to Bretton Woods institutions, and the obvious buy-in of both the warring parties and of numerous UN agencies and international NGOs, the Office of the UNR/HC was ultimately in a position to override consensual decision making to give priority to the restructuring of the overall Sudan operation. The very considerable autonomy of the UNR/HC and the lack of a clear accountability structure meant that NMPACT was very vulnerable to changes in priorities and policy from the top.

A further change of leadership in late 2004 – both with the NMPACT coordinator and the UNR/HC – has allowed the programme to refocus on its original objectives and the principles of engagement and to rebuild its partnerships with national counterparts and institutions. In March 2005 the Coordination Structure carried out a review to examine the continued relevance of NMPACT in a post-peace scenario and to analyse ways in which the programme can readjust its goals and principles in order to contribute to the implementation of the CPA.

The post-peace scenario: Reinventing NMPACT to support the implementation of the CPA

The Third Partners' Forum, which was held in February 2005, focused on reassessing the continued role of NMPACT in a post-peace scenario. The forum concluded that the NMPACT framework, its goal and its principles of engagement continued to be highly relevant to the current regional context. The partners felt that the emphasis on 'conflict transformation' in the approach of the overall programme framework remained relevant, if not critical, in a post-CPA era (Office of the UNSRSG and R/HC, 2005).

There was widespread concern amongst the NMPACT partners, including the official counterparts, that the protocols making up the CPA had not addressed all of the root causes of the conflict. However, the partners believed that underlying issues that could lead to renewed tension had to be tackled through democratic, non-violent means by the local community and that the NMPACT model could be instrumental in fostering dialogue and constructive interaction in the region. The Third Partners' Forum affirmed the commitment of the NMPACT partners to a renewed effort to focus on the principles of engagement, particularly on the principle of fostering an enabling environment for an indigenous, Nuba-led long-term peace process, which remains essential in this phase. The forum also concluded that NMPACT partners should focus on supporting successful power sharing between the warring parties and the integration of the two administrative entities, the Nuba Mountains (the old South Kordofan state) and West Kordofan state, which have been merged into a new, enlarged State of South Kordofan according to the provisions of the CPA (Office of the UNSRSG and R/HC, 2005). This required an official clarification or amendment to the original programme document by the two counterparts as the NMPACT mandate is currently restricted to the areas covered by the Burgenstock Cease-fire Agreement, i.e. the five provinces of today's South Kordofan and only Lagawa Province in West Kordofan.

Given the special conditions accorded to the Nuba Mountains by the Two Areas Protocol signed in Naivasha in May 2004 and endorsed within the CPA in January 2005 and the general dissatisfaction of many Nuba about the agreement[5] (cf. Nuba Survival, 2005), failing to successfully implement the CPA in the new South Kordofan state may pose a challenge not just for the reconstituted state, but for the entire CPA in the country as a whole (Office of the UNSRSG and R/HC, 2005). In this regard, it is important that the spirit and the principles of NMPACT be retained in any new humanitarian and development intervention and in the coordination of the aid efforts that will have to be redesigned to reflect the change of context in the Nuba Mountains. Drawing on the experience of NMPACT, any new arrangement should be built on an analysis on how to best support the implementation of the protocol, including the merger of the state institutions and the engagement of the Missirya communities of West Kordofan, a large pastoralist group belonging to the Baqqara Arab tribe, with Nuba groups in the state.

Conclusions and lessons learned for policy and practice in complex emergencies

Coordination in complex emergencies

The experience of NMPACT and the processes that led up to it, albeit short, offer significant lessons for programming in complex emergencies, be it in other areas of Sudan or in countries with a similar context. NMPACT was developed out of learning from the OLS experience and capitalized on the shortcomings of that response to bring about changes that were unprecedented in the history of humanitarian engagement in Sudan. In particular, NMPACT set out to bring a long-term perspective into an emergency context through its focus on the principles of engagement and its emphasis on national ownership, participatory development as related to programme design and decision making and collective advocacy. The strong inter-agency coordination around the principles allowed the programme to break with the pattern of traditional externally driven responses to food insecurity and to adopt an approach focused on capacity building, promotion of sustainable agriculture and market revitalization alongside conflict transformation and peace building.

Coordination in crisis contexts is traditionally difficult to achieve. Agencies' focus on visibility, competition for funds and an excessive attention to organizational self-interest (emphasis on own mandate rather than the interests of the intended beneficiaries) means that often coordination has little appeal in humanitarian contexts. Furthermore, in acute emergencies the humanitarian sector tends to privilege speed over quality of assistance and there is a fear that coordination would cause unnecessary delays (Van Brabant, 1999). In this regard, agencies do not consider that emergencies often become protracted and therefore the most effective responses are not necessarily the speediest ones. NMPACT's experience has shown that it is important to learn lessons that can help plan for the medium and long-term while the crisis is still ongoing. The research work on land tenure issues, which was carried out while the conflict was still active, has been crucial in informing the peace process and today is providing a sound basis for external interventions aimed at supporting IDPs' return and agricultural rehabilitation in the region.

In complex emergencies, agencies are also reluctant to create another 'layer of bureaucracy', so the challenge is to make coordination effective. This usually requires a cost, as effective coordination is time and staff intensive and needs to be properly resourced (Van Brabant, 1999). Again, the lessons learnt from NMPACT are that in the absence of an adequately staffed coordination structure the effectiveness of the programme was much reduced, the focus on the principles was weakened and, more importantly, the sustained interaction between the warring parties, which was a crucial element of success of the model, was severed, with the unwelcome effect of hindering the feasibility of cross-line operations for the partners.

Van Brabant (1999) argues that in order for coordination to be effective, it needs to fulfil a number of functions, which range from serving as a contact point to providing situational updates, fulfilling security, learning and training functions as well as performing functions related to programming, political analysis, representation and strategic decision making. Table 3.6 summarizes the main functions performed by the NMPACT Coordination Structure.

The model of coordination offered by NMPACT was uncontroversial because it focused on providing services to partners and facilitating learning and analysis, rather than assuming a strong lead role in decision making or management of security issues. The principles of engagement were originally designed to prevent the Coordination Structure from focussing on day to day management of the operation on the ground, something some of the partners were reluctant to accept. The emphasis of the Coordination Structure was therefore shifted to exercising quality control of the operation and supporting the partners in their endeavour to be true to the principles. The donors' support for NMPACT was also undoubtedly another important factor that made the framework appealing to some of the partners.

Table 3.7 Key functions of NMPACT coordination structure

Key functions	*Details*
Services to members	• Venues for cross-line meetings • Salary surveys and labour legislation • Maps
Information	• Collective agency contact point/agency directory • Facilitation of information flow • Lead baseline assessment
Situational updates	• Produce situational updates • Monitor and collate needs assessments and surveys • Monitor and collate resource availability
Security	• Information exchange on security situation
Learning/ evaluation	• Collect programme reports/reviews • Identify research and commission studies (e.g. on land and environment) • Interagency discussion of reviews/evaluation • Carry out reviews/evaluations • Develop institutional memory of lessons identified
Programming	• Database of projects (sectors/area) • Sectoral policies/guidelines • Facilitation of interagency programme planning and cross-line programming • Review programming gaps/duplication • Operational role to fill gaps
Political analysis	• Conflict analysis • Agency position in the political economy of the conflict • Scenario development • Mediation and confidence building between HAC and SRRC and between them and the agencies
Representation	• To powerbrokers to negotiate framework of consent and access to humanitarian space • To donors for resource mobilization • To ceasefire monitoring mission and political actors for advocacy
Strategic decision-making	• About agency position in the conflict and principles of engagement

Source: Adapted from Van Brabant (1999)

NMPACT's experience shows that there is much to gain from strategic coordination in complex emergencies, when analysis, discussion, monitoring and review of the situation and ongoing and planned interventions are required.

NMPACT and innovation: The principles of engagement and political humanitarianism

The focus of the principles of engagement on sustainability, equitability and 'do least harm' pushed for a shift in emphasis within NMPACT away from short-term emergency intervention and externally driven aid delivery. The medium- to long-term focus of NMPACT's food security intervention has proven to be effective in enhancing the potential for recovery and building the resilience of local communities in the Nuba Mountains. The findings of the twin-track analysis presented in this study document the change in trend from emergency interventions to longer term responses over the three years of life of NMPACT. Preliminary observations from peer reviewers on the impact of the NMPACT partners' interventions seem to indicate that NMPACT's approach to food security, with its emphasis on advocacy to remove barriers to sustainable livelihoods security, including through collective advocacy to obtain a cease-fire and a monitoring body, had an important role in terms of strengthening people's own capacities to enhance their resilience and lower their dependence on external food aid.

The NMPACT framework was also successful in using aid to foster dialogue between the warring parties. The adoption of the 'do least harm' approach resulted in joint advocacy to end the humanitarian blockade and to press for a cease-fire. The response was characterized by extensive engagement with the GoS, the SPLM, key diplomatic players and the cease-fire monitoring operation. The so-called 'political humanitarianism' of NMPACT can be looked upon as a model to address livelihoods issues in a complex emergency by focusing on responses based on political analysis, advocacy, fostering links with key actors in the political and peace-keeping spheres of operation, and strong local ownership of the recovery process. The significant results achieved by NMPACT in a relatively short space of time indicate that much can be learned from a response that is informed by a political analysis of food insecurity and entitlements deprivation, which departs from the more conventional technical and community-centred responses of aid agencies to such crises.

Much remains to be tested and understood in the context of programming in complex political emergencies. NMPACT's experience, while of a short duration, shows that there is a clear role for applying long-term and systematic development thinking to emergencies and supporting learning and analysis of the deep-rooted causes of the main elements of a crisis to generate informed responses. While the need for quick external aid delivery cannot be avoided in the event of major crises or emergencies, there is definitely a need to adopt and adapt alternative models in contexts where such emergencies have become chronic and where there are political elements that need to be tackled to unblock the crisis. Its relevance for Sudan is particularly high at a moment when peace and confidence building are very much on the agenda and when the situation in Darfur risks becoming a chronic emergency, where the international response is strongly driven by the provision of external inputs and needs to further invest in understanding local political and livelihoods' realities to inform interventions; realities to inform interventions.

References

AACM International (1993) *Land Use Survey*, prepared for the Southern Kordofan Agricultural Development Project, AACM, Adelaide.

African Rights (1995) *Facing Genocide: the Nuba of the Sudan*, African Rights, London.

African Rights (1997) *Food and Power in Sudan – A Critique of Humanitarianism*, African Rights, London.

Ajawin, A. and A. de Waal (eds.) (2002) *When Peace Comes. Civil Society and Development in Sudan*, Red Sea Press, Asmara.

Alden Wily, L. (2005) Guidelines for the Securitization of Customary Land Rights in Southern Kordofan and Blue Nile States, Sudan. Customary Land Security Project USDA/USAID PASA, Washington.

Anderson, M. (1999) *Do No Harm – How Can Aid Support Peace – Or War*, Lynne Rienner Publishers, London.

CARE (2002) *Nuba Mountains Programme Development. Field Assessment, Draft Report. Part I*, prepared by John Plastow and Nancy Balfour, CARE RMU, Nairobi

Flint, J. (2001) 'Democracy in a war zone: The Nuba parliament'. In S. M. Rahhal (ed.) *The Right To Be Nuba. The Story of a Sudanese People's Struggle for Survival*, Red Sea Press, Asmara, pp. 103–112.

Harragin, S. (2003a) *Desk Study on Land-Use Issues in the Nuba Mountains, Sudan*, Background Report to Accompany the Literature Review/Annotated Bibliography, Concern Worldwide and Save the Children US, Nairobi.

Harragin, S. (2003b) *Annotated Bibliography on Land-Use Issues in the Nuba Mountains, Sudan*, Concern Worldwide and Save the Children US, Nairobi.

Harragin, S. (2003c) *Nuba Mountains Land and Natural Resources Study, Main Report*, 15 December 2003 draft, NMPACT and USAID, Khartoum/Nairobi.

IFAD (2000) *Sudan South Kordofan Rural Development Programme. Appraisal Report*, Main Report, Appendices and Annexes, Draft (Version 1.0), IFDA, Rome.

IFAD (2004a) *South Kordofan Range Management Strategy Study and Khor Abu Habil Catchments Basin Planning and Water Development Study*, Final Report, Volume 1: Main Report – South Kordofan Range Management Strategy Study, YAM Consultancy and Development and GIBB Africa, Khartoum.

IFAD (2004b) *South Kordofan Range Management Strategy Study and Khor Abu Habil Catchments Basin Planning and Water Development Study*, Final Report, Volume 3: Appendices, YAM Consultancy and Development and GIBB Africa, Khartoum.

IFAD (2004c) *South Kordofan Range Management Strategy Study and Khor Abu Habil Catchments Basin Planning and Water Development Study*, Final Report, Annexes: Appendix D, Volume 3: Water Resources, YAM Consultancy and Development and GIBB Africa, Khartoum.

IOM/UNDP (2003) *Sudan IDP Demographic, Socio-Economic Profiles for Return and Reintegration Planning Activities*, Nuba IDP Households, IOM/UNDP, Khartoum

Johnson, D. (2003) *The Root Causes of Sudan's Civil Wars*, James Currey, Oxford.

Karim, A., M. Duffield, S. Jaspars, A. Benini, J. Macrae, M. Bradbury, D. Johnson, G. Larbi and B. Hendrie (1996) *OLS: Operation Lifeline Sudan: A review*, University of Birmingham, Birmingham.

LTTF (Land Tenure Task Force) (1986) *Strategy for Development of Rainfed Agriculture*, Annex II, Land Tenure Task Force Main Report, Khartoum University Press, Khartoum.

Macrae, J. and Leader, N. (2000) *Shifting Sands: The Search for 'Coherence' between Political and Humanitarian Responses to Complex Emergencies*, HPG Report 8, ODI, London.

Manger, L. (1994) *From the Mountains to the Plains. The Integration of the Lafofa Nuba into Sudanese Society*, Nordiska Afrikainstitutet, Uppsala.

Manger, L., Pantuliano, S. and Tanner, V. (2003a) *The Issue of Land in the Nuba Mountains*, Office of the UN Resident and Humanitarian Co-ordinator for the Sudan, Khartoum.

Manger, L., Egemi, O., El Tom El Imam, A. and Pantuliano, S. (2003b) *Options Available for Dealing with Land Tenure Issues in the Nuba Mountains*, Office of the UN Resident and Humanitarian Co-ordinator for the Sudan, Khartoum.

Nadel, S. F. (1947) *The Nuba. An Anthropological Study of the Hill Tribes of Kordofan*, Oxford University Press, Oxford.

NFSWG (Nuba Food Security Working Group) (2001) Food Security Assessment and Intervention Strategy, Nuba Mountains, Southern Kordofan. Internal Document, Nairobi.

Nuba Survival (2005) *Nuba Marginalized by Naivasha Peace Process*, Press Statement, 4 March, Nuba Survival, London.

Office of the UNSRSG and UNR/HC (UN Special Representative of the Secretary-General and UN Resident and Humanitarian Co-ordinator) in the Sudan (2005) Third NMPACT Partners' Forum Report, United Nations, Khartoum.

Office of the UNR/HC (UN Resident and Humanitarian Co-ordinator) in the Sudan (2002a) Rapid Needs Assessment of the Nuba Mountains Region, United Nations, Khartoum.

Office of the UN R/HC in the Sudan (2002b) *Programme Framework Document. Nuba Mountains Programme Advancing Conflict Transformation (NMPACT)*, Khartoum: United Nations.

Office of the UNR/HC in the Sudan (2002c) *NMPACT Information Table*, United Nations, Khartoum.

Office of the UNR/HC in the Sudan (2002d) *First NMPACT Partners' Forum Report*, United Nations, Khartoum.

Office of the UNR/HC in the Sudan (2002e) *Second NMPACT Partners' Forum Report*, Khartoum: United Nations.

Office of the UNR/HC in the Sudan (2002f) *Report of the Baseline Data Collection Exercise for the NMPACT programme – Summary Findings, Nuba Mountains Region*, United Nations, Khartoum.

Office of the UNR/HC in the Sudan (2003) *NMPACT Internal Review: March 2002 – July 2003*, United Nations, Khartoum.

Office of the UNR/HC for the Sudan (2004a) *Sudan Transition and Recovery Database (STARBASE), SPLM Controlled Nuba Mountains*, Version 2, United Nations, Nairobi.

Office of the UNR/HC for the Sudan (2004b) *Sudan Transition and Recovery Database (STARBASE), South Kordofan State*, Version 2, United Nations, Nairobi.

Pantuliano, S. (2003) *Harnessing the Potential of Aid to Protect Livelihoods and Promote Peace – the Experience of the Nuba Mountains Programme Advancing Conflict Transformation (NMPACT)*, Proceedings of the FAO International Workshop on Food Security in Complex Emergencies, Tivoli (Rome), 23–25 September, p. 14.

Pantuliano, S. (2004) *Understanding Conflict in the Sudan: An Overview*, The World Bank Group, Washington DC.

Pantuliano, S. (2005) 'Changes and Potential Resilience of Food Systems in the Nuba Mountains Conflict', http://www.fao.org/docrep/008/af141e/af141e00.htm.

Pingali, P., Alinovi, L. and Sutton, J. (2005) 'Food security in complex emergencies: Enhancing food system resilience', *Disasters* 29 (1): S5–S24.

Rahhal, S. M. (2001) 'Focus on crisis in the Nuba Mountains'. In S. M. Rahhal (ed.) *The Right To Be Nuba. The Story of a Sudanese People's Struggle for Survival*, Red Sea Press, Asmara, pp. 36–55.

Saeed, A. A. R. (2001) 'The Nuba'. In Rahhal, S.M. (ed.) *The Right To Be Nuba. The Story of a Sudanese People's Struggle for Survival*, Red Sea Press, Asmara, pp. 6–20.

Salih, M. A. M. (1984) 'Local markets in Moroland: The shifting strategies of the Jellaba merchants'. In L. Manger (ed.) *Trade and Traders in the Sudan*, Department of Social Anthropology, Bergen, pp. 189–212.

Salih, M. A. M. (1995) 'Resistance and response: Ethnocide and genocide in the Nuba Mountains, Sudan', *Geo-Journal* 36 (1): 71–78.

Shazali, S. (2004) *National Human Development Report*, Draft (internal document), UNDP, Khartoum.

SKRPU (South Kordofan Rural Planning Unit) (1980a) Nuba Mountains Agricultural Production Corporation (NMAPC) Technical Report, SKRPU, Kadugli.

SKRPU (1980b) *NMAPC (Nuba Mountains Agricultural Production Corporation) Technical Report. Annex 1: Soils and Vegetation*, SKRPU, Kadugli.

SKRPU (1980c) *NMAPC Technical Report. Annex 2: Water Resources*, SKRPU, Kadugli.

SKRPU (1980d) *NMAPC Technical Report. Annex 3: Livestock Production. Annex 4: Crop Production*, SKRPU, Kadugli.

SKRPU (1980e) *NMAPC Technical Report. Annex 5: Population, Social Organisation and Production Systems*, SKRPU, Kadugli.

SKRPU (1980f) *NMAPC Technical Report. Annex 6: Economics and Marketing*, SKRPU, Kadugli.

SKRPU (1980g) *NMAPC Technical Report. Annex 7: Land Use and Planning Regions*, SKRPU, Kadugli.

Stevenson, R. (1984) *The Nuba People of Kordofan Province: An Ethnographic Survey.* Monograph 7, Graduate College Publications, Khartoum.

Suliman, M. (1997) 'Ethnicity from Perception to Cause of Violent Conflict: The Case of the Fur and Nuba Conflicts in Western Sudan', a Contribution to CONTICI International Workshop, Bern: 8–11 July 1997, Institute for African Alternatives, London.

UNCERO (1999) *Report of an Inter-Agency Assessment Mission to the Nuba Mountains of South Kordofan, Sudan*, United Nations, Khartoum.

UNDP (1996) Area Rehabilitation Scheme, Kadugli SUD/95/004: Report on a Visit to Kadugli by Project Formulation Mission, UNDP, Khartoum.

UNFPA/CBS (Central Bureau of Statistics) (2003) *Data Sheet for Sudan by States*, UNFPA/CBS, Khartoum.

UNICEF/AET (2003) *School Baseline Assessment Database southern Sudan*, UNICEF, Nairobi.

United Nations (1999) *Presentation on South Kurdufan State*. Special Focus on the Nuba for the UN Nuba Mountains Mission (19– 27 June 1999), United Nations, Khartoum.

Van Brabant, K. (1999) *Opening the Black Box. An Outline of a Framework to Understand, Promote and Evaluate Humanitarian Co-ordination*, paper commissioned by the Disaster Studies Programme of the Centre for Rural Development Sociology, University of Wagenigen, The Netherlands.

White, S. (2003) *Environmental Guidelines and Screening Criteria for Project Planning. Developed for Use by NMPACT and its Development Partners in the Nuba Mountains Region of South Sudan*, internal document, Nairobi.

Wily, L. A. (2004) *Suggested Inputs on the Subject of Land Ownership in the Implementation Modalities for the Interim Period*, internal document, Nairobi.

World Bank (2003) *Sudan. Stabilisation and Reconstruction. Country Economic Memorandum*, The World Bank Group, Washington DC.

World Bank (2004) *Nuba Mountains Community Empowerment Project*, Development Grant Facility (DGF)/Post Conflict Fund (PCF), The World Bank Group, Washington DC.

CHAPTER 4

Policies, practice and participation in protracted crises: The case of livestock interventions in southern Sudan

Andy Catley, Tim Leyland and Suzan Bishop

Abstract

This chapter describes the importance of livestock as a fundamental livelihoods asset for agropastoral communities in South Sudan, and explains the complex linkages between the ownership and use of livestock, conflict, marketing systems, seasonality and vulnerability. It also examines how programmes to support pastoralist livelihoods in southern Sudan have introduced innovative, participatory elements that go beyond the traditional humanitarian framework. In particular, it tracks livestock interventions in the Operation Lifeline Sudan (Southern Sector) Programme since the early 1990s.

Introduction

Southern Sudan is renowned as a place where people revere livestock and where livestock, particularly cattle, play a fundamental role in human food security. The longevity and severity of Sudan's civil war had devastating impacts on livestock, among many other aspects of life, until the January 2005 CPA between the GoS and the SPLA established South Sudan as a region with a semi-autonomous government. In the preceding 15 years, international donors had supported livestock interventions consistently as a means to improve food security. Since the early 1990s this support was delivered mainly through a network of NGOs within OLS, initially under the coordination of the United Nations Children's Fund (UNICEF) and later under the FAO. The OLS Livestock Programme not only successfully controlled rinderpest – a cattle disease of major local and international importance – but also developed a primary animal health system based largely on developmental rather than relief principles (Tunbridge, 2005). The general philosophy of the programme, especially until the late 1990s, was that despite the operational challenges of implementing field activities in southern Sudan and short timeframes for projects, something more than relief was possible and necessary. Rather than regarding local people as recipients of relief aid designed and delivered by outsiders, the developmental approach used community participation and cost recovery to lay a foundation for the long-term provision of livestock services. The programme evolved within the overall emergency focus of OLS. It challenged typical humanitarian thinking, which stressed the importance of neutrality and impartiality and the rapid delivery of free inputs.

This chapter reviews the development of community-based animal health worker (CAHW) services and rinderpest control in southern Sudan within the context of severe, protracted crisis and repeated back-to-back cycles of relief funding. It examines the factors that enabled an extensive community-based system to evolve and the processes through which programme-wide operational guidelines and policies were created and adjusted. Central to this analysis are issues of programme coordination and evidence-based approaches to decision making. To analyse coordination issues, the chapter draws heavily on an extensive review of United Nations coordination efforts in emergencies (Donini, 1996) that identified three broad categories of coordination: 'coordination by command', based on strong leadership endorsed by authority and often involving some control of resources for implementing agencies; 'coordination by consensus', which harmonizes and encourages responses in relation to common objectives, in the absence of formal authority; and 'coordination by default', primarily a process of information exchange between actors in the absence of a formal coordination entity. The chapter looks in detail at the style of coordination in the OLS Livestock Programme using these three types of coordination as a framework for analysis, and describes how and why coordination differed under UNICEF and FAO. In terms of evidence-based approaches to programming and policies, it also examines ways in which information was generated and shared within the programme, and specifically, the value of this information in relation to assessment of programme impact.

The chapter is divided into four main sections. The first provides an overview of the livelihoods of the main livestock-rearing communities in southern Sudan, focussing on the linkages between livestock, conflict and food security. It then describes the OLS Livestock Programme between 1990 and 2004, including the introduction of developmental approaches and related coordination issues; the section includes some important historical perspectives and an account of the various policy actors, their narratives and their interactions both within and outside the programme. The next section is a technical assessment of information in the programme with a focus on impact, addressing the extent to which evidence of impact was collected and used. The final section draws out the overall lessons learned and makes recommendations for improving the design, coordination and policy support to livelihoods-based interventions in protracted crises.

Livestock and livelihoods in southern Sudan

The social and economic importance of livestock

The main livestock-rearing communities in southern Sudan can be broadly categorized as pastoralist and agropastoralist. The pastoralists include the Toposa, Jie, Murle and Nyangatom, who occupy relatively dry, lowland areas in Eastern Equatoria. The agropastoralists include the Dinka, Mundari and Nuer, who occupy the flood plains of Bahr el Ghazal, Lakes, Jonglei and Upper Nile. A second, smaller cluster of agropastoralists lives in the hills of Eastern Equatoria. Although these communities access and use a variety of assets, the ownership of cattle is the common and most prevalent livelihood strategy. Estimates of livestock populations vary, with most attention given to numbers of cattle. Jones (2001) estimated 6.8 to 7.8 million cattle in the whole of southern Sudan.

The value of livestock in southern Sudanese pastoral and agropastoral groups relates to their contribution to food, income, agricultural production, kinship ties and marriage. Both agropastoralists and pastoralists keep mixed herds of cattle, sheep and goats that provide meat, milk, manure, as well as hides and skins, which can be exchanged or sold for grain. Chickens are also kept, particularly by poorer Dinka households. Crop production is more important to the agropastoral groups (see Table 4.1) although the Dinka, Nuer and Toposa all grow sorghum and other crops such as sesame and millet. Fishing is important to both the Dinka and Nuer.

In southern Sudan social obligations and interactions revolve around cattle. The desire to acquire cattle is a major factor in how people behave; social cohesion is based on the exchange of cattle at marriage, and involves complex systems of loans, gifts and co-ownership or joint decision making about cattle between kinfolk and friends. Among the Dinka and Nuer, daughters are prized as future sources of cattle to be acquired through marriage, and

Table 4.1 Food economies of pastoralists and agropastoralists in southern Sudan

	Ethnic group, animal husbandry type, location					
	Dinka, agropastoral, Bahr el Ghazal		Nuer, agropastoral, Central Upper Nile		Toposa, pastoral, Eastern Equatoria	
Food item	*Normal year food economy (%)*	*Households with few cattle (%)*	*Normal year food economy (%)*	*Households with few cattle (%)*	*Normal year food economy (%)*	*Households with few cattle (%)*
Milk	25	5	30	15	20	20
Meat[1]	5	0	10	10	45	45
Exchange	15	5	20	10	5	10
Fish	20	15	10	20	–	–
Wild food	10	20	5	10	5	5
Crops	25	55	25	35	25	10
Gift	–	–	–	–	–	10

Note: [1]Categorized as meat and blood for the Toposa.
Source: Adapted from Fielding et al. (2000)

young unmarried men strive to acquire cattle as soon as possible in order to marry. Deng (1987) provides a detailed explanation of the way in which Dinka families are founded on bride wealth and explains how the exchange of cattle at marriage extends far beyond an economic transaction into the core of Dinka kinship relationships.

It used to be that livestock were often sold at livestock auctions, which were a feature of southern Sudanese market towns since before independence. However, as conflict and insecurity cut off the supply of goods and services for purchase, these auctions were curtailed and livestock markets collapsed. The supply of goods and services gradually began to increase again in the mid-1990s when NGOs started to bring in commodities such as soap and salt to pay their workers, and these items often ended up in markets. As border and road access slowly improved, second-hand clothes and consumables found ready markets and the livestock auctions began to reappear.

A major livestock trade route was for cattle to be walked from Bahr el Ghazal and Lakes to northern Uganda, and then trucked to slaughter. A smaller route existed for animals walked from Toposa areas of Eastern Equatoria to Narus, and then trucked through Lokichokio to Nairobi for slaughter. Small numbers of animals would also be walked northwards to Kordofan and Darfur from Bahr El Ghazal, and to Malakal and Ethiopia from the Sobat Basin. The war forced these trade routes to operate informally and prevented the establishment of roads, communication and marketing infrastructures, which might have enabled trade.

Vulnerability and food insecurity

For decades, the dominant problem in southern Sudan was the pursuit by the GoS and SPLA of a violent resolution to the civil war. Although these opposing groups were signatories to the tripartite agreement that allowed OLS to exist, persistent conflict hindered access to vulnerable communities and repeatedly undermined or destroyed the assets of people in them. In the context of relentless insecurity, meaningful investment in, or protection of, any material asset was very difficult. Government development policies in the north were largely irrelevant in the south, even in GoS-controlled areas. Economic policies focusing on the export of oil and livestock were commonly expressed in the south as violent interventions by GoS forces and militias.

At the macro level, the international community allowed war and atrocities to continue. For decades, foreign governments, the UN and African regional bodies such as the Organization of African Unity and the InterGovernmental Authority on Development lacked either the will or the capacity for meaningful facilitation of conflict resolution. The combination of complex regional politics linked to the Middle East, key resources in Sudan such as oil and water, and GoS political guile led to a chronic conflict that was often overlooked on the international scene. Approximately 1 million people died and up to 3 million people were displaced in southern Sudan during the early and mid-1980s. It was not until 250,000 people died in 1988 that OLS was created through a UN agreement with the GoS and SPLA/M. Although OLS was a milestone in the history of humanitarian intervention and kept large numbers of southern Sudanese alive, by early 2005 southern Sudan had some of the worst poverty indicators anywhere in the world. The New Sudan Centre for Statistics and Evaluation estimated that 90 per cent of the population lived on less than one US dollar a day, primary school enrolment was the lowest in the world, infant mortality was 150 out of 1,000 live births and under 5 mortality was 250 out of 1,000 (NSCSE, 2004).

Types of conflict: War and raiding

Severe food deficits and famine among cattle-rearing communities in southern Sudan have often arisen due to a combination of chronic conflict, repeated shocks such as droughts, crop pests and animal disease epidemics, and seasonal variations in food availability. The onset of civil war in Sudan in 1956 and the alignment of the Dinka with the southern rebels led to counter-insurgency warfare based on raids by Arab Rizeigat and Misseriya militia into Dinka territory, which were repeated over more than 25 years. The early 1990s saw increased GoS military expenditure, greater commitment to counter-insurgency and adoption of scorched-earth tactics (Deng, 1999; 2002). In his detailed analysis of the famine in Bahr el Ghazal in 1998 in which 100,000 people died, Deng (1999) explains the interplay between historical trends, drought and the depletion of herds and interruption of planting due to severe conflict between pro-GoS Dinka militia and the SPLA. He also discusses the failure of markets and traditional redistribution mechanisms to cope with very large-scale livestock and grain losses. The response of the SPLM and OLS was very inadequate. The Bahr el Ghazal famine was unusual because conflict-related livestock losses were partly attributable to fighting within the Dinka community rather than between the Dinka and other ethnic groups. In this situation, wealthier households with more cattle were more susceptible to violence because their attackers had intimate knowledge of cattle ownership and distribution (Deng, 2002).

Adding to the violence and vulnerability associated with war was the practice of cattle raiding between neighbouring communities in the south. In the 1980s and 1990s, inter-ethnic livestock raiding between the Nuer and Dinka became a serious problem, as did cross-border raids between the Toposa and the Turkana (living in northwest Kenya). These raids were particularly violent and involved the theft of such large numbers of animals that whole communities were left destitute. Large areas of potential grazing land became inaccessible areas, traditional communication between groups collapsed and politicians and militias incited revenge attacks by increasingly well-armed youths.

The situation was compounded in the late 1990s by the GoS drive to access oil beneath the Nuer grazing lands of Western and Eastern Upper Nile. In order to avoid putting foreign oil workers at risk, a scorched earth policy was used to shift southern rebel sympathizers away from oil wells, pump stations and pipelines. It was relatively easy to arm militias to carry out the policy because youths had become accustomed to war and conflict, and traditional leadership over them was waning. Young men were also willing to carry out revenge attacks to recover raided cattle. The GoS approach was very effective and was followed up with spurious development projects by the oil companies, keen to win the support of people who were left behind (Coalition for International Justice, 2006).

The conflict, raiding and land clearances were devastating in terms of aid delivery. As we describe later in the chapter, the OLS Livestock Programme coped better than most sectors because it had invested in raising the awareness of livestock owners and CAHWs and the programme moved with the people.

Seasonality and shocks

A normal year for southern Sudanese cattle herders is characterized by marked seasonal variations in food availability. Milk supply depends on the time of calving and available and accessible grazing resources and water. Most herders manage livestock breeding so that births coincide with good pasture. This means that offspring are born during the wet season or early dry season, and that milk (for both calves and people) is most available from around May to August. Crop production is also planned according to the wet season, with planting in April and May, and harvesting in July and August (but sometimes also as late in the year as October). As the dry season progresses from November onwards, supplies of grain and milk begin to decline and people are more reliant on wild foods, blood from cattle (sometimes mixed with milk) and the sale or exchange of livestock for grain. These seasonal trends result in a regular, annual period of human nutritional stress commonly called 'the hunger gap', which in pastoralist areas occurs towards the end of the dry season and into the main wet season. Similar seasonal patterns of food availability are evident in agropastoral areas although relatively good grazing and milk supply can be maintained until the rains. Children, teenagers and the few adults who accompany the cattle to the cattle camps rely on little more than milk, fish and wild foods for up to four months.

Variable rainfall in the southern Sudanese lowlands means that on average, one in three crop plantings fail. Risk of crop failure is further increased by conflict (having to flee before harvest), bird damage, disease and attacks by army worms. Crop failures can be remarkably localized, with one district having poor harvests while neighbouring districts have surplus production. This vulnerability to shortage of food grains has encouraged the development of sophisticated exchange mechanisms both within pastoralist communities and with neighbouring agrarian tribes in higher-rainfall areas. Exchange of livestock (particularly cattle) for grains remains strong and occurs mainly during the pre-harvest hunger gap and following harvest.

Dependency on livestock and crops makes the southern Sudanese vulnerable to epidemics of animal and crop diseases. In livestock, rinderpest was the most important disease affecting food security. It is an acute and highly contagious viral disease that affects primarily cattle, but also some species of wildlife. The disease was introduced into Africa in the late 1880s when cattle were the main source of wealth in rural African communities, for everyone from settled farmers to pastoralists. By the early 1890s it had spread throughout the continent and killed 90 to 95 per cent of the cattle population.

The southern Sudanese have various names for rinderpest, which reflect the devastating impact the disease used to have on their cattle and livelihoods. The Dinka Bor call it *nyan tek*, meaning 'one calf remains'. When CAHW services were first introduced into southern Sudan under UNICEF in 1993, rinderpest was reported almost daily (Riek Gai Kok, cited in Tunbridge, 2005). Although rinderpest was thought to have been eradicated from southern Sudan in 2006, many other diseases affect livestock and the livelihoods of their owners, including foot and mouth disease, contagious bovine pleuropneumonia, trypanosomosis and fasciolosis, among others.

Livestock interventions and policy process

When reviewing livestock interventions in southern Sudan and the formal or informal policies that influenced those interventions, it is obvious that the overriding institutional issue affecting people in the south during the period under study was the GoS and SPLM commitment to a violent resolution to the civil war. Despite both sides being signatories to the tripartite agreement that allowed OLS to exist, conflict persisted, hindering access to vulnerable communities and repeatedly undermining or destroying their assets. In the context of such relentless insecurity, meaningful investment in, or protection of, any material asset was difficult to achieve. Nevertheless, lessons can be learned from the Livestock Programme as a long-term relief intervention that adopted a developmental approach. The following section focuses on livestock interventions from before 1989 to early 2005, and in particular, the development of CAHW services and rinderpest control.

The pre-OLS period (before 1989)

> Rinderpest eradication is a waste of time in southern Sudan. We can only do 'fire engine' livestock interventions and these must all be free of charge. Involving local people compromises our neutrality and in any case, what do pastoralists know? (Typical statements from international livestock agencies, NGOs and the SPLA)

Up to independence in 1956 the development of livestock policies and services in Sudan followed a pattern similar to many other African countries. Colonial administrations invested heavily in livestock services legislation, government departments, vaccines and diagnostic laboratories, and veterinary schools appeared in the early to mid-1900s (Jack, 1961). However, services were always relatively weak in the south and the onset of war exacerbated the deficit. Policies and services in the north continued to influence the south, but only in the transition zone (between north and south) and in government-held areas.

The main policy actors in the period immediately before OLS were donors, NGOs and the SPLA. While donors provided funding, the SPLA played a dominant role in deciding how projects were implemented on the ground. Heavily influenced by the Marxist ideologies of the Mengistu regime in neighbouring Ethiopia, the SPLA approach was to support free veterinary inputs in the areas they controlled.

For the international community and southern Sudanese in the early 1990s, rinderpest eradication was a priority. The disease was killing large numbers of cattle in southern Sudan but the Pan African Rinderpest Campaign of the Organization of African Unity was struggling to prevent or control outbreaks, mainly because insecurity prevented access to cattle-keeping communities. The Global Rinderpest Eradication Programme (GREP) of FAO realized the importance of the Sudan problem both for Africa and globally, but reports from FAO and the Organization of African Unity from this period indicate a certain pessimism about progress towards rinderpest eradication.

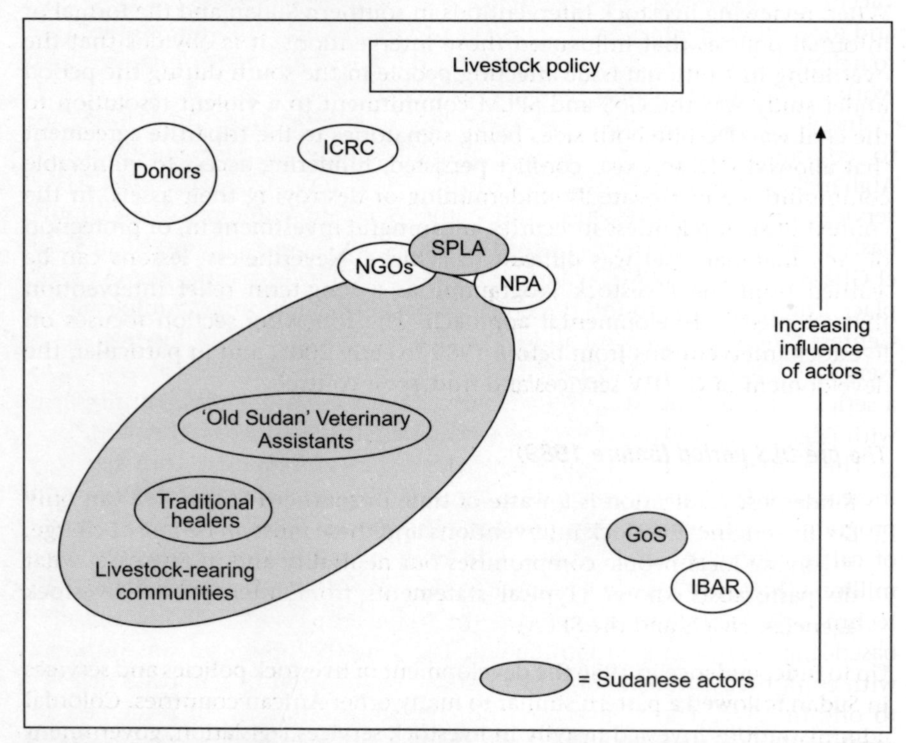

Figure 4.1 Changing policy actors and linkages over time: Before 1989

Support to CAHW projects in southern Sudan began in the mid-1980s when Oxfam supported a para-vet project with the Action Committee for the Promotion of Local Initiatives in Self-Help, a local NGO in Eastern Equatoria (Almond, 1987). In common with many other CAHW projects run by NGOs in Africa up to the late 1990s, the project was effective but small-scale and isolated from policy debates at national or subnational levels. Concepts of community participation, indigenous knowledge and local capacity to deliver were still largely unheard of within the veterinary establishment in Sudan or among international livestock agencies.

Scaling up community-based delivery systems (1993 to 2000)

> Rinderpest eradication is possible using community-based approaches. We must involve local players and introduce systems related to future sustainability. (Typical statements from UNICEF, NGOs, African Union/InterAfrican Bureau for Animal Resources (AU/IBAR), SPLM, FAO/GREP)

When OLS was created in 1989, livestock interventions were part of the UNICEF household food security programme. Working with Oxfam, Norwegian People's Aid and ICRC, UNICEF supported rinderpest control based on conventional vaccination approaches, using a series of refrigeration facilities (a cold-chain) and formally trained workers to deliver vaccines to remote communities. Due to heightened conflict, by the end of 1992 the ICRC had pulled out of livestock work and the UNICEF programme had stagnated.

In 1993 a technical adviser was seconded to UNICEF from Tufts University to coordinate the livestock programme. Having worked previously in Afghanistan, the adviser had assisted the UNDP and NGOs to set up CAHW services and was familiar with participatory development approaches. On the basis of that experience, participatory assessments were conducted in Bahr el Ghazal and Upper Nile, where rinderpest was rife. The initial surveys took place against a background of severe human malnutrition: according to the Centers for Disease Control and Prevention, 80 per cent of children under 5 were critically undernourished. Discussions with community leaders led to a series of social contracts. Herders were trained as vaccinators and supplied with heat-stable rinderpest vaccine (which did not require a cold-chain).

The community-based approach achieved results quickly, with ten fold increases in rinderpest vaccination figures. Whereas prior to 1993 the OLS programme was vaccinating around 140,000 cattle per year, the number of cattle vaccinated in 1993 was 1.48 million, in 1994 the figure was 1.74 million and in 1995 it was 1.07 million (Leyland, 1996). In 1996 a FAO/GREP technical consultation commended the 'pioneering work' and community-based approaches in southern Sudan. The renewed optimism was justified, with confirmed outbreaks of rinderpest decreasing from 11 outbreaks in 1993 to one in 1997. There have been no confirmed outbreaks of rinderpest in southern Sudan since 1998; in 2002, the GoS was prepared to declare Sudan provisionally free of the disease on a zonal basis (Jones et al, 2003).

At that time, the programme focused on working with traditional local institutions and their leaders – the cattle camp leaders, chiefs and subchiefs – who were influential and able to mobilize and organize livestock keepers. The cattle camps were existing, well-organized and well-managed groups of livestock keepers under traditional leadership, which represented the ideal entry point for the programme. The philosophy and approach of the programme was warmly received on the ground by community leaders (Leyland, 1996). The SPLM attitude towards rinderpest eradication and CAHWs became more supportive. Many SPLM leaders and commanders were cattle owners who recognized the impact of rinderpest control achieved using CAHWs.

By early 1993 it became clear that a broader community-based animal health programme was feasible in southern Sudan. Livestock keepers were already noticing the impact of rinderpest vaccination and wanted other diseases to be controlled. In May 1993, the first OLS Livestock Coordination Meeting was convened with NGOs, the Sudan Relief and Rehabilitation Association and the Relief Association of Southern Sudan, and it was agreed that the programme should expand to include vaccinations and treatments for other livestock diseases. By 1994 the Office for Foreign Disaster Assistance (OFDA) of USAID and the European Commission's Humanitarian Aid Office (ECHO) were funding a wider CAHW programme covering rinderpest and other diseases. With the prospect of funding commitments from these donors for the sector, UNICEF invited NGOs to join the programme and set up CAHW projects in under-served areas.

The initial period of CAHW services under UNICEF were characterized by strong coordination of increasing numbers of NGOs and trained Sudanese animal health workers within OLS. Although only UNICEF and Oxfam-GB were involved in the programme in 1993, by 2002 there were 12 NGOs working under UNICEF coordination.[1] The programme used 1,500 CAHWs under the supervision of 150 local veterinary supervisors and coordinators, and 40 NGO field veterinarians and livestock officers (Jones et al, 2003). A standardized training manual for CAHWs was adopted by the programme in 1997, and minimum standards and guidelines in 1999. The training manual included a standardized CAHW training curriculum agreed upon with all NGOs and other partners. In addition to this example of a coordinated approach to the CAHW service, other approaches also evolved within the programme that contrasted sharply with the conventional relief style of OLS at the time. These approaches included payment for services, the employment of southern Sudanese, capacity-building initiatives and the establishment of training centres in southern Sudan to provide higher-level technical supervision of CAHWs, as summarized in Box 4.1.

An important feature of UNICEF coordination was the agency's control of key programme resources such as veterinary vaccines and medicines. Implementing NGOs were required to adhere to collectively agreed operating principles such as cost recovery, and to follow programme-wide methodologies such as those relating to the design of CAHW systems. At the same time, NGOs

Box 4.1 Developmental approaches in protracted crises: Elements of the community-based animal health system in southern Sudan

Community participation

- Participatory assessment of animal health problems
- Use of indigenous livestock skills and knowledge
- Use of traditional leadership, such as cattle camp leaders, as entry points
- Local selection of CAHW trainees
- Some involvement in evaluation and impact assessment

Payment for services

- Early recognition that CAHWs working on a voluntary basis lacked sufficient incentives to continue working
- Participatory assessment indicated that livestock keepers were willing to pay for basic veterinary care
- Mixed success on the ground but the long-term perspective prevailed, which assumed that if people had been paying for clinical services during the war, the move to 'full privatization' would be relatively easy after the war

Employing southern Sudanese in the programme

- Challenged the humanitarian principle of 'neutrality' and a strict OLS policy of not employing Sudanese
- Local Sudanese translators and trainers viewed as essential to the livestock programme
- Following recruitment of a Dinka translator by the programme, gradually became normal practice for all UN agencies and NGOs to employ southern Sudanese in all types of posts

Local capacity-building: support to Veterinary Coordination Committees and Sudanese NGOs

- Support was provided to community-level Veterinary Coordination Committees (VCCs) for awareness raising and mobilization, management of drugs and cost recovery funds, supervision of animal health workers and planning livestock activities with CAHWs.
- Mixed success on the ground: by early 2005 the VCCs were no longer functional in most areas, or had been dissolved and re-formed in an attempt to become more representative.
- In 1998 the first southern Sudanese NGO dedicated to livestock work was established and joined the OLS consortium. Initially called Vetwork Sudan and later Vetwork Services Trust, it received considerable material and moral support from UNICEF and AU/IBAR, and capacity-building support from Christian Aid, PACT and other agencies.

Training centres for Animal Health Auxiliaries in South Sudan

- Animal health auxiliaries received four months of training and were designed to improve supervision and support to CAHWs.
- Two Southern Sudan Animal Health Auxiliary Training Institutes were set up by VSF-B, one in Marial Lou, Tonj County in 1996 and the other in Mankien, Upper Nile in 1997. Due to insecurity, the Mankien centre was later moved to Mading, but there, too, fighting related to livestock raiding forced its closure.
- In 2004 the Marial Lou institute was absorbed into the New Sudan Livestock Training Centre under the Livestock Training Centre Act 2004 (Laws of New Sudan); this effectively marked the handover of the centre from an NGO to South Sudan government ownership and control.

Sources: Adapted from Leyland, 1996; Jones et al., 2003; Bishop, 2003.

recognized the benefits of UNICEF coordination in terms of the technical support and direction provided by the coordination team, which included practical inputs such as training NGO veterinarians on the ground. All these features of coordination resulted in a relatively strong process for developing programme policies that were generally supported and implemented by southern Sudanese actors, NGOs, AU/IBAR and FAO/GREP (see Figure 4.2).

A shift in coordination methodology after 2000

By 2000 the security situation had improved in many areas of southern Sudan, although the Upper Nile remained highly problematic. In general, access to communities became easier and NGO staff spent more time on the ground talking to people and reviewing experiences. Interest in local capacity-building was revived through various means. As markets began to open up, trade improved and options for working with the private sector became clearer. By 2005 the SPLM Directorate of Animal Resources and Fisheries had been established, and three southern Sudanese veterinarians – all with direct experience of CAHW systems – were appointed to key positions related to livestock development. With such changes taking place, it was an opportune

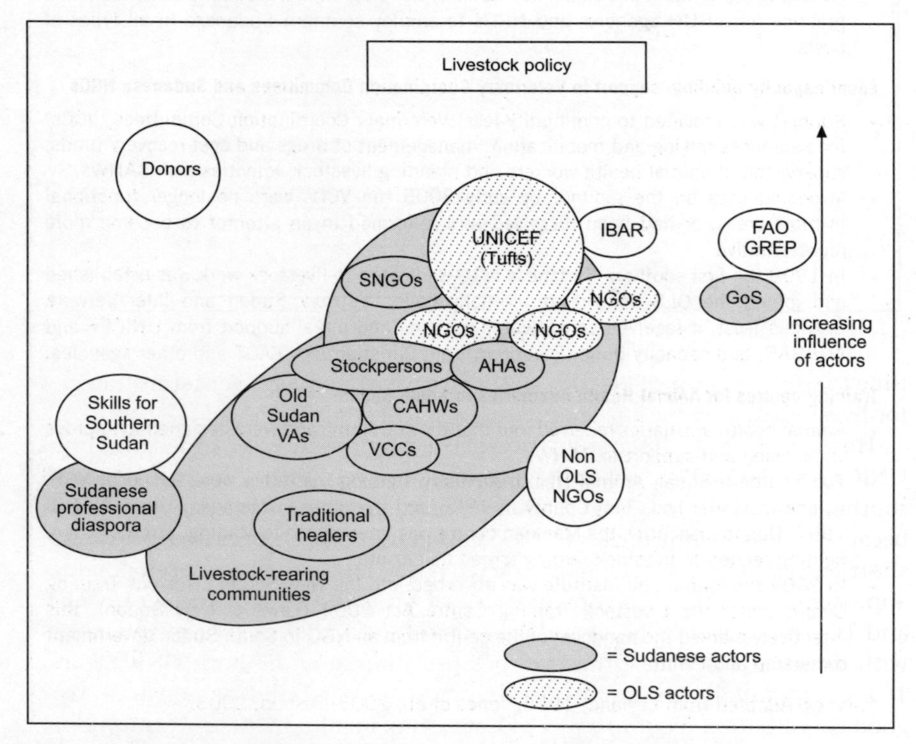

Figure 4.2 Changing policy actors and linkages over time: UNICEF–OLS 1999

Box 4.2 Coordination and policy process in the OLS Livestock Programme under UNICEF and FAO: Perceptions of NGO and UN practitioners and programme managers

As part of the research for this chapter, ten senior NGO and UN practitioners and programme managers were identified, all with experience of the OLS Livestock Programme under both UNICEF and FAO coordination. These informants were asked, independently, to score the two agencies against 20 indicators of coordination and policy process (0=lowest score to 10=highest score). The informants were then asked to rank the 20 indicators in order of importance (1st = most important to 20th = least important). The median duration of field experience in southern Sudan for the 10 informants was six years.

The median scores (95 per cent confidence intervals) for the 10 most highly ranked indicators are shown below. The median rank of each indicator is shown in parentheses.

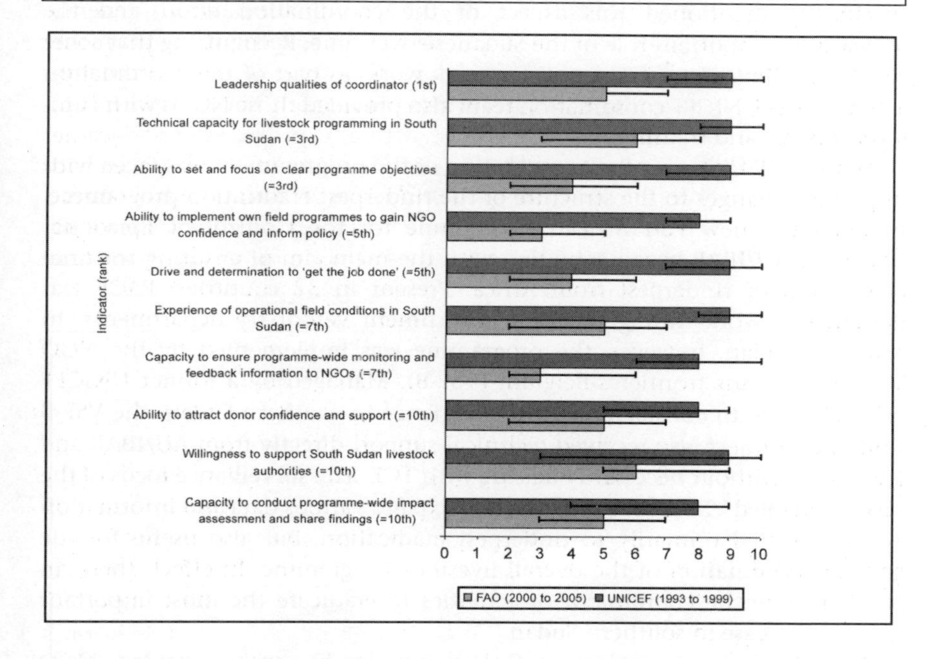

time to work with the SPLM and NGOs to develop some longer-term strategies for livestock services in the south.

However, in 2000 the coordination of the programme was handed over from UNICEF to FAO's TCE. NGO programme managers with long experience in southern Sudan viewed the TCE takeover of the coordination role with concern because it was assumed the TCE operational approach focussed on typical short-term relief work, even in protracted crises, rather than on developmental approaches. In order to compare coordination methodologies under UNICEF and TCE we consulted NGO and UN practitioners and programme managers with direct experience of the livestock programme under both UNICEF and TCE coordination. The results are shown in Box 4.2 and demonstrate that good coordination requires a mix of leadership, management and professional qualities, plus capacity to use field-level engagement to attain the confidence of partners and inform policy dialogue. The indicator referring to technical capacity (ranked equal third by informants) covers the developmental approach of the programme under UNICEF relative to the more relief-focussed approach of TCE.

A review of the dynamics and direction of the livestock coordination meetings under UNICEF indicates that programme policies evolved largely based on clear programme objectives, and personal interaction and negotiation between UNICEF staff, NGOs and SPLM. While the coordination meetings were initially dominated by expatriate and Kenyan vets, as more southern Sudanese workers gained experience they became more involved. Informants specifically mentioned this aspect of the coordination effort and the increasingly important role of the Sudanese over time. Recognizing that some NGOs had limited experience of livestock work, as part of the coordination activities the UNICEF coordination team also provided those NGOs with field-level support and training.

When TCE took over the coordination of the programme it was faced with important changes to the structure of the rinderpest eradication programme. In 2000 the new Pan African Programme for the Control of Epizootics (PACE) of AU/IBAR began activities, with the main aim of ensuring the final eradication of rinderpest from Africa. Present in 32 countries, PACE was designed to work through African government veterinary departments. In southern Sudan, however, the programme was implemented by the NGO Vétérinaires sans frontières-Belgium (VSF-B). Managed by a former UNICEF coordinator with considerable field experience in southern Sudan, the VSF-B rinderpest project also received technical support directly from AU/IBAR and FAO/GREP without necessarily liaising with TCE. The surveillance focus of the project enabled VSF-B to strengthen and expand various data and information systems related primarily to rinderpest eradication, but also useful for the general coordination of the overall livestock programme. In effect, then, an NGO became the coordinator of activities to eradicate the most important livestock disease in southern Sudan.

As TCE began to implement CAHW projects in southern Sudan, NGO fears about their approach and capacity were realized. Both NGO and TCE informants noted how funding constraints and bureaucracy led to weak field implementation, and in some areas, over-reliance on NGO resources. The changing nature of coordination under TCE relative to the previous coordination under UNICEF is exemplified by the production of NGO training manuals in 2004 and 2005 that were initiatives of individual NGOs rather than a programme-wide effort. This was an indication that NGOs had started to operate independently of the wider programme. The overall policy process was becoming more fragmented relative to the period under UNICEF coordination between 1993 and 1999 (Figure 4.3).

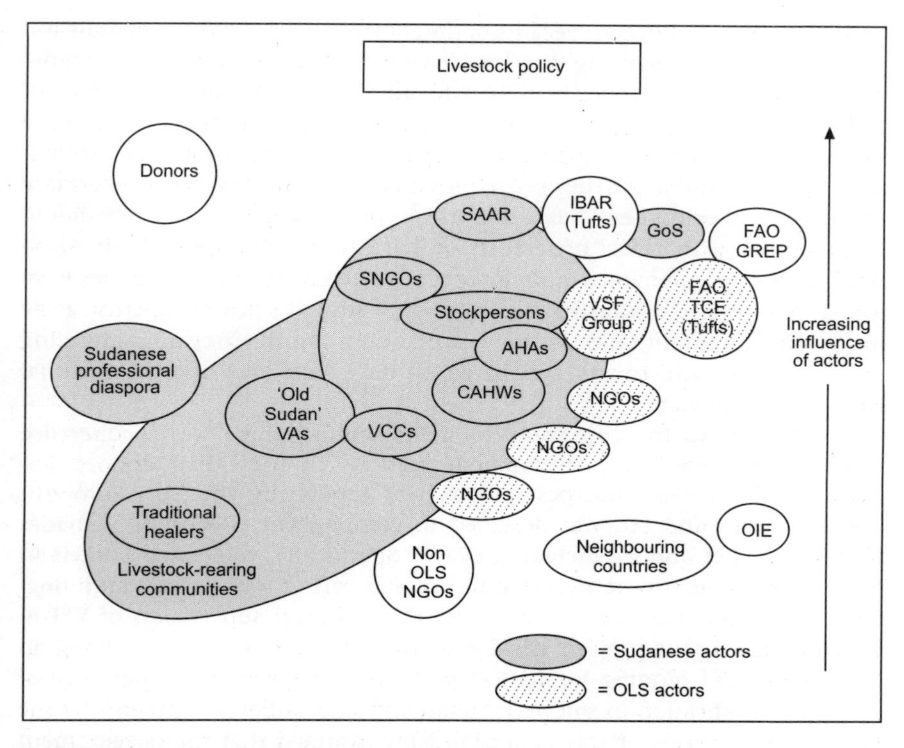

Figure 4.3 Changing policy actors and linkages over time: Early 2005

The donor influence: Supportive flexibility versus bureaucracy and incoherence

The developmental approaches initiated by UNICEF in the livestock programme would not have occurred without considerable flexibility on the part of relief donors. In the case of OFDA the principles of self-sufficiency, enhanced recovery, participation and strengthening local capacity consistently guided its support to the programme. Some donors during different periods formed strong ties with the programme, with real donor interest in programme successes and problems, and an understanding of the realities on the ground. With the exception of OFDA, however, lessons learned remained with individuals and were not incorporated into donor policies or guidelines. As those individuals moved on, new staff arrived with limited knowledge of southern Sudan and limited awareness of developmental thinking in chronic emergencies.

Although relief donors were sometimes able to push and even overstep the boundaries of conventional relief funding, the developmental approach of the livestock programme was seriously constrained by short-term funding cycles and donor bureaucracy. Short funding cycles consumed huge amounts of

UNICEF and NGO staff time because of the constant need to write proposals and produce reports on existing projects. Developmental thinking was hindered because longer-term strategies were difficult to formulate in the absence of committed funding. While developmental approaches require more staff time for facilitation, technical support to local partners and organizational learning, relief funding emphasizes the need to provide material inputs such as veterinary medicines and equipment. Some donors specified how project budgets should be structured, with a clear preference for 'hard' over 'soft' inputs. At the same time, donor procurement requirements for materials such as medicines were sometimes intensely bureaucratic, thereby diverting technical expertise away from more important tasks. In some cases, donor priorities regarding funding (of 'hard' over 'soft' inputs) was based on unwritten rules and the personal whims of donor staff.

In addition to the constraints of relief donors, there was incoherence between the relief and development sections of individual donors. For example, the VSF-B rinderpest project was funded by the EU's European Development Fund but was designed to complement NGO field activities funded by relief donors, including ECHO. Specifically, veterinary workers in NGO relief projects were expected to become part of a large-scale and long-term disease surveillance system under the technical supervision of VSF-B. When designing the project, VSF-B recognized there was a risk in relying on continued relief funding for project partners, but given the importance of rinderpest eradication in southern Sudan and substantial investments by the EU over many years, it was understandably assumed that the development and relief sections of the EU would liaise to ensure that sufficient funds were indeed guaranteed to NGOs working with the project. However, by mid-2003 ECHO was reviewing its support to the livestock programme and proposing (verbally) 20–50 per cent cuts in support to NGO CAHW projects.

As another example of incoherence between a donor's relief and development sections, in 2000 the DFID support to southern Sudan focussed on conventional relief inputs, due in part to concerns at London headquarters about the need for neutrality and impartiality in the south, and owing to long-running debates about the definitions of 'relief' and 'rehabilitation'. Funding was largely restricted to short-term material inputs. At the same time, DFID committed funding to a four-year development project in AU/IBAR that aimed at facilitating supportive policy and institutional arrangements for community-based animal healthcare in the Greater Horn of Africa region – including southern Sudan. Initially called the 'Community-based Animal Health and Participatory Epidemiology' project, the regional project supported various activities in southern Sudan related to policy, such as studies on veterinary privatization, livestock marketing and capacity-building. Furthermore, by early 2005 the project was also engaging the SPLM directly and providing technical support to it to develop policy frameworks for the livestock sector. While the message from the relief wing of DFID avoided notions of sustainability or engagement of local actors, the message from the development wing was rather different. These examples of policy incoherence reflect the frequently reported policy and structural divisions between the relief and development departments of major donors (Harmer and Macrae, 2004).

National, regional and global trends

The emergence of the livestock programme in southern Sudan did not occur in isolation from national and international trends, particularly in relation to supportive policies for CAHW-type approaches. Although often resented by the veterinary establishment in various countries in the Horn of Africa region, by early 2005 AU/IBAR had worked with government veterinary services in neighbouring countries to the north, east and south to develop varying levels of legislation to support CAHWs. Kenya, Uganda, Ethiopia and northern Sudan also established community animal health units in their departments of veterinary services. These units were tasked with the quality control and coordination of CAHWs and the promotion of private veterinary facilities linking CAHWs to veterinarians or diploma holders.

At the global level, AU/IBAR also worked with the World Organisation for Animal Health (OIE) to revise the international standards on veterinary services to include CAHWs as one type of veterinary para-professional (Catley et al, 2005). The OIE is mandated by the World Trade Organization to set global animal health standards as they relate to international trade under the Agreement on the Application of Sanitary and Phytosanitary Measures. Changes to the OIE Code effectively created global acceptance of CAHWs according to international standards. In Africa, AU/IBAR published its Africa-wide policy on CAHWs in 2002.

Impact, information and policy

The OLS Livestock Programme generated a huge number of reports, including quarterly and annual progress reports, evaluations, impact assessments, reviews and specific studies. Agencies involved included UNICEF, NGOs, FAO, AU/IBAR and various donors, universities, consultancy companies and individuals. Studies were conducted on livestock marketing, veterinary privatization, local capacity-building, livestock diseases and other topics. Databases were also created, initially in UNICEF and later in FAO. This section of the chapter examines this substantial body of information from the perspective of policy process. In particular it aims to identify evidence of programme impact and show the relationship between this evidence and policy development or change within the programme. In terms of impact assessment, the programme produced three main types of evidence: rinderpest-specific reports; general project and programme monitoring and evaluation reports; and community participatory evaluations.

Rinderpest vaccination and surveillance

The OIE oversees the eradication of major livestock diseases and sets standards and guidelines for the formal recognition of freedom from rinderpest on a country basis. The eradication process follows the 'rinderpest eradication pathway', with various stages of the pathway having clearly defined targets based on quantified epidemiological indicators and procedures. The pathway for rinderpest eradication in southern Sudan involves two main stages: mass vaccination (1989 to June 2002) and surveillance (from June 2002 onward). Mass vaccination was intended to reduce the circulation of rinderpest virus to a minimal level; indicators of progress included reduced clinical disease (fewer disease outbreaks). The vaccination stage was followed by a period of zero vaccination, watching to see if the disease reappeared. The early surveillance period aimed to detect any new outbreaks as rapidly as possible and respond with localized vaccination to eliminate the final remnants of virus. Later stages of surveillance involve carefully designed surveys to detect evidence of rinderpest infection, and includes blood sampling cattle and testing the blood for rinderpest antibodies.

The introduction of CAHWs and heat-stable rinderpest vaccine in 1993 was associated with a 10.6-fold increase in vaccination figures and a steady reduction in rinderpest outbreaks. Sudan is now considered to be provisionally free of the disease. The comprehensive rinderpest surveillance system serves to maintain the benefits of freedom from disease and respond rapidly to any new outbreaks (detailed surveillance indicators and measures of progress are available in Jones et al, 2003.)

Given the impact of rinderpest on food security in southern Sudan, the absence of rinderpest is a reasonable proxy indicator for the overall livestock programme. Based on cattle population and rinderpest mortality estimates, the cost–benefit of rinderpest control in southern Sudan was estimated at 1:13 (Blakeway, 1995). When compared with the provision of food aid using delivery costs and nutritional benefits, the cost–benefit ratio of rinderpest control versus food aid was 34:1 (see Box 4.3).

Routine project monitoring and evaluation

The largest body of information on the livestock programme comprises general NGO project monitoring and evaluation reports, including monthly, quarterly and annual reports depending on donor requirements. With up to 12 NGOs involved in the programme since the mid-1990s, around 1,400 reports were produced over a 10-year period. However, almost all of these reports focussed on the measurement of *process* (the delivery of inputs and training), rather than of *impact* on human livelihoods or food security. This fixation with process measurement was related to three main constraints. First, donor reporting formats emphasized the use of process monitoring indicators such as 'number of people trained' or 'number of animals treated'. These reporting systems

Box 4.3 The cost–benefit of rinderpest control in southern Sudan

Short-term benefits through reduced rinderpest mortality

- Rinderpest mortality is highest in younger cattle, less than 3 years of age. Around 60 per cent of this age group will die during an outbreak, and outbreaks occur every two to four years (average three years).
- Using proportional piling to assess herd age structures, around 38 per cent of cattle were less than 3 years old.
- Assuming a cattle population in the south of 4 million (1995 estimates), 304,000 cattle would die from rinderpest each year. Market value of young cattle was US$25 per head.

Based on these assumptions and a control programme that protected up to 50 per cent of the total cattle population, the immediate savings was US$3.8 million per year.

Long-term benefits through increased production; relation to food aid

- In the absence of rinderpest, the cattle population would grow at around 7 per cent per year. Proportional piling indicated that an average herd composition of 40 per cent adult cows. A cow calves every two years and produces about 1.3 litres of milk per day for human consumption for most of the year.
- Based on these assumptions, rinderpest control would produce 72,800 litres of milk per day.
- Three litres of milk provides the daily calorie needs of one adult person. Other livestock-derived foods (direct and indirect) contribute another 75 per cent of food needs relative to milk. Rinderpest is controlled in only 50 per cent of the cattle.
- Based on these assumptions, cattle-derived foods after rinderpest control would feed 21,000 people.
- The World Food Programme (WFP) provides 400 g of cereal per person per day. Food aid provision in southern Sudan costs US$1,000 per metric ton (mt).
- Based on these assumptions, 21,000 people would require 3,001 mt of cereal costing US$3 million (the total food aid delivered to southern Sudan in 1994 was 21,844 mt). The combined short- and long-term benefits of rinderpest control amount to US$6.8 million.

Cost–benefit ratio of rinderpest control compared with food aid

- The total cost of the UNICEF-OLS Livestock Programme in 1994 was US$500,000, of which US$200,000 was for hard inputs (vaccine and vaccination equipment); assume benefits to 21,000 people as described above. The hard inputs of food aid to provide the same food benefits to 21,000 people costs US$3 million.
- The cost–benefit ratio of rinderpest control is 0.5/6.8 = 1:13.
- If staff and logistical costs are ignored, the input cost ratio for rinderpest control versus food aid to achieve the same level of benefit is 0.2/(3.0 +3.8) = 1:34.

Source: Blakeway, 1995

reflected the overall focus of relief agencies on the delivery of inputs and the need for financial accountability. Second, agencies and individuals within the programme interpreted 'impact' in different ways. Third, programme impact on human food security impact was perceived as difficult to measure due to the complex set of production and social benefits of livestock, the various effects on these benefits of different diseases in various species and the difficulty of defining how to attribute preventive or clinical interventions relative to other factors such as improved access to pasture or water.

In general, NGOs associated impact with veterinary service provision and therefore, they measured numbers of veterinary workers, their distributions and their activities. Within this definition of impact, the most commonly reported information was livestock vaccination and treatment figures, and there was an assumption that the prevention or treatment of livestock diseases resulted in improved human food security. Although NGOs, UNICEF, FAO and donors seemed to rely heavily on these figures as evidence of programme success, the figures are useful for impact measurement only if they are combined with livestock population data and estimates of disease prevalence by disease and livestock species.

Community participatory evaluation

Participatory approaches to project evaluation were introduced into the OLS Livestock Programme by VSF-B and VSF-Switzerland in 1999 (Catley, 2000). The basic assumption was that if communities were making important contributions to programme design and implementation, then they should also be involved in programme evaluation. The approach also recognized the indigenous knowledge of livestock keepers in southern Sudan, and their powers of observation regarding animal health and production. The initial adaptation and testing of participatory evaluation methods drew heavily on the work of ActionAid-Somaliland (ActionAid, 1994) and the International Institute for Environment and Development (Guijt, 1998), but also used standardized methods to enable statistical analysis. An outline of the methodology and some results are presented in Box 4.4.

Over time, community participatory evaluation was used by Save the Children-UK in Bahr el Ghazal and Upper Nile (Okoth, 2001) and by the VSF agencies in Western Upper Nile (Hopkins and Short, 2001), Bahr el Ghazal (Hopkins and Short, 2002; Hopkins, 2003) and Shilluk Kingdom (Hopkins, 2002). These assessments provide consistent results on the positive livelihoods impact, which was attributable to the activities of CAHWs and other veterinary workers in the programme. Depending on location, impact on livelihoods included more animals, more milk for consumption and sale, improved child nutritional status, increased bride wealth prices, opportunities for restocking through exchange and increased social status. Relative to the mass of routine project monitoring and evaluation reports mentioned above, this handful of participatory assessments was highly influential in terms of adaptations actually made to NGO interventions.

Information flow and policy dialogue

Southern Sudan is characterized by severe operational constraints, with very weak infrastructure and huge distances between the OLS base in Lokichokio, northern Kenya, and field locations. During the first 10 years of the OLS

Box 4.4 Community participatory evaluation in the OLS Livestock Programme: Links between human food security and animal health interventions

Community Participatory Evaluation involved four main stages:

1. Define the spatial and temporal boundaries of the project using methods such as participatory mapping and time-lines, respectively
2. Describe the benefits derived from livestock and their relative importance
3. Describe changes in animal health during the project period
4. Describe project attribution in relation to project benefits and problems

Results were cross-checked against secondary data such as vaccination and treatment figures by disease type.

Example of results: benefits derived from livestock and their relative importance (results derived from proportional piling of community-defined benefits with five informant groups)

Benefit	*Median score*
Milk	34
Dowry	24
Dung	10
Compensation/fines	9
Income	7
Meat	6
Butter	3
Ploughing	3
Hides and skins	2
Ceremonies	1

Example of results: changing livestock disease prevalence attributed to a community-based animal health project, 1996–1999

N=6 informant groups, W = 0.61, $p<0.01$

Diseases (Nuer–English):
Gieng – rinderpest
Liei – mixed parasitism
Rut – haemorrhagic septicaemia
Doop – CBPP
Dat – FMD
Duny – ephemeral fever
Yieth ping – sudden death

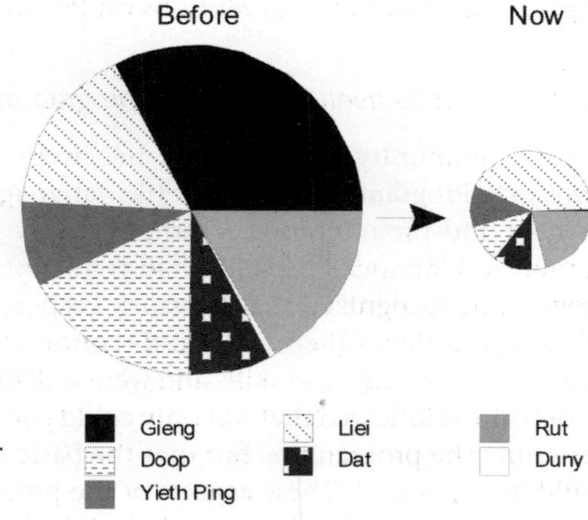

This example indicates a dramatic reduction in gieng (rinderpest) during the three years of the project. To some extent, this result was verified by reference to rinderpest vaccination figures.

Livestock Programme, communication was limited to radio contact, meetings between people and written reports. Reports were written by hand, data was summarized using pocket calculators and reports were hand-carried by pilots or project staff from the field. It was a verbal and paper-based system, before the days of satellite phones and laptop computers.

Despite these limitations, for many years there was a strong flow of information within the programme. The OLS livestock coordination meetings were a key event for information flow and learning. For NGO vets who were new to community-based approaches, the coordination meetings were a chance to seek advice from others, plan activities and gain assurance that many other people were experiencing the same logistical and security problems as they were. A simple reporting system was in operation and the coordination meetings were the mechanism for NGO staff to present information on their work and receive guidance from UNICEF. The meetings provided an important policy forum for the programme, both in terms of discussion on policy issues during the meetings and more informal information-sharing and lobbying. After many years there was still active debate on issues such as the role of local veterinary coordination committees (VCCs) (see Box 4.1) and cost recovery mechanisms, and it was evident that despite the protracted crisis context, most of these issues were related more to development than to relief.

Conclusions

Coordination and innovation: Factors for success

Experiences from the OLS Livestock Programme in southern Sudan demonstrate that livelihoods-based food security programming is possible in protracted crises. It requires commitment to livelihoods approaches, a strong but flexible coordination effort with control over resources and support to systematic impact assessment of interventions on livelihoods.

Commitment to livelihoods-based programming

When community-based approaches were first introduced into the OLS Livestock Programme in 1993, the language and concepts of livelihoods analysis and programming – terms such as 'assets', 'social capital', 'human capital' and 'financial capital' – had yet to appear in aid organizations. Yet the programme recognized that livestock keepers in southern Sudan were highly reliant on cattle for their social and economic well being, had rich indigenous livestock knowledge and skills and were well organized, with strong traditional leadership. It followed that not only could communities be involved actively in designing the programme, but that the basic elements of a sustainable service could be supported. These aspects of the programme might now be described as 'participatory', 'developmental' or 'livelihoods-based', but they contrasted sharply with the typical emergency interventions in OLS at the time – and yet proved to be highly successful. The lesson learned in this case is that in protracted crises, relief agencies have much to learn from the participatory approaches and livelihoods thinking of development agencies.

Strong but flexible coordination

Under the UNICEF Livestock Programme, coordination was characterized by command and consensus. *Command* was made possible by UNICEF controlling the supply of veterinary vaccines and medicines to implementing NGOs, and strong technical direction leading NGOs to seek advice and support from the coordination team. It was also made possible by UNICEF being the technical focal point in southern Sudan for the Pan African Rinderpest Campaign, and the need for agencies involved in rinderpest control to follow common, programme-wide strategies. The *consensus* element of UNICEF coordination was exemplified by the regular programme coordination meetings during which technical issues were raised and discussed and consensus was reached on programme strategies. Although full consensus among all actors was rarely reached on issues such as cost recovery or VCCs, a sufficient level of agreement was reached for adopting programme-wide approaches. In their assessments of the coordination effort, NGO programme managers and practitioners prioritized a mix of leadership, technical skills and management capacities (see Box 4.2). The lesson learned was that the technical capacity of the coordinating body will not carry much weight unless it is combined with good management and communication.

Although the need for strong UN coordination in protracted crises has been discussed at length (Donini, 1996; Minear, 2002), effective coordination was especially necessary in the OLS Livestock Programme due to the untested CAHW approach being advocated by UNICEF, and the developmental nature of the programme. When CAHWs were first introduced in the early 1990s, the international veterinary establishment was somewhat sceptical. Organizations such as FAO and AU/IBAR were struggling to control rinderpest using conventional approaches and while they recognized that southern Sudan was a particular challenge, the notion of illiterate (but trained) herders vaccinating cattle was difficult for them to accept. But the strong coordination of this novel approach in southern Sudan ensured that vaccination coverage improved dramatically and rinderpest outbreaks declined. In less than three years, FAO and AU/IBAR were convinced that community-based approaches were indeed a useful adjunct to rinderpest eradication in conflict-affected areas. Since the mid-1990s, both agencies have been consistent supporters of CAHWs both for rinderpest control and for more general primary veterinary service delivery. In one sense, the southern Sudan experience had a global impact in terms of the policies of international agencies. When viewed from a food security perspective, the eradication of rinderpest from southern Sudan might be regarded as the single most important achievement for protecting key livelihoods assets and supporting food security in the area.

The broader, more general developmental approaches promoted by the livestock programme in southern Sudan also required strong coordination. The overriding character of OLS was determined by humanitarian responses, which were constrained by short-term programming and for many agencies, a strict adherence to the humanitarian principle of 'neutrality', which prevented the participation of Sudanese. Within this general programming environment, the Livestock Programme aimed to achieve more than relief and used developmental approaches very early in its history. While other technical sectors of OLS initially regarded the Livestock Programme's approach as inappropriate, the impact of the Livestock Programme was soon recognized. By the late 1990s other sectors, particularly human health, were liaising with the programme to use CAHWs to deliver non-livestock inputs.

Impact assessment and policy

In development circles, the emergence of livelihoods frameworks has improved understanding of the policy and institutional constraints affecting the acquisition and management of assets by poor or vulnerable communities. Related to this trend has been increased attention to the processes through which new policy evolves, whether in developing or industrialized countries. Conventionally, policy change or development has often been viewed as a logical, linear process in which new information is collected in relation to a policy problem, and policy makers automatically use this information to change policy. This view of policy process as objective and value-free is particularly appealing to natural scientists because it follows a logic similar to that used in designing much scientific research.

However, studies on policy process challenge this linear view of policy and instead highlight the chaotic and value-driven nature of policy research and dialogue (Keeley and Scoones, 2003). They argue that policy change is driven by dominant policy narratives and that research and information, however rigorous, are always interpreted according to personal incentives and past experiences. This research also noted that practice often precedes policy and that because information is always incomplete, policy decisions necessarily involve risk.

The more chaotic definition of policy process fits well with experiences in the OLS Livestock Programme. When a community-based approach to livestock services was advocated in the early 1990s, there was no hard evidence that this approach would work in southern Sudan. The decision to use CAHWs and adopt more developmental approaches was based almost entirely on professional judgements by the team involved and information derived from participatory assessments with livestock-rearing communities. When the idea of community-based approaches to rinderpest control was first presented to donors such as OFDA, the approach made intuitive sense to those advisers who had direct field experience in southern Sudan, despite the lack of scientific evidence. At the same time, it was important that there was recognition that conventional approaches to rinderpest control had failed even though rinderpest control was acknowledged as both an international public good and a priority food security intervention in southern Sudan.

Over time, experiential, field-level learning by the coordination team continued to inform the technical direction of the programme and gave the team members credibility to lead because they understood the challenges of working on the ground. Although this learning rarely involved conventional surveys or studies, it did use quantitative indicators of service provision and rinderpest eradication to track progress. The programme coordination meetings provided the space for policy dialogue and for NGO practitioners to contribute to programme development, guidelines and operating procedures. Policies were negotiated and moved forward when a critical mass of professional opinion was reached on a particular issue. When specific policies later proved to be difficult to implement, a system for reviewing experiences and holding dialogue was already in place.

Before 1999 agencies in the OLS Livestock Programme tended to regard impact assessment in isolation from project design and implementation, and assessments were most often conducted by external consultants. Although communities contributed to identifying problems and implementing projects, their views were often sidelined during impact assessment. The introduction of community participatory evaluation by some NGOs helped to bridge a gap between the need for better information on programme impact and the need to involve local people in assessing change. In the future, community participatory evaluation methodology could be improved by more triangulation of information, measurement of attribution and repetition of standardized methods with representative samples. However, the extent to which community participatory evaluation becomes institutionalized will also depend in part on donor reporting requirements and in particular on whether there are changes in the current fixation on process rather than impact. It is inappropriate for NGOs to spend so much time and effort collecting and submitting data on process when it fulfils little more than a bureaucratic function. This does not mean that information on process is of no value, but that a far more appropriate balance between process, impact and organizational learning needs to evolve in the context of protracted crises.

South Sudan and future livestock policy process

Following the signing of the CPA in January 2005, it was appropriate to hand over the control of livestock policy formulation in southern Sudan from a UN coordinating body to the new Government of South Sudan. In early 2005 AU/IBAR began supporting the Secretariat of Agricultural and Animal Resources to draft a policy framework for the livestock sector and this process involved NGOs, FAO TCE and AU/IBAR together with SAAR personnel. However, a range of major policy and capacity issues still needed to be addressed.

In 2003 AU/IBAR consulted senior policy makers in East Africa to gather their views on policy and institutional constraints affecting the livestock sector (AU/IBAR, 2004). Ministers of agriculture, permanent secretaries and heads of livestock departments were interviewed in Ethiopia, Kenya, Sudan (north), Tanzania and Uganda. The key findings from this consultation included a realization among policy makers that they often lacked knowledge on how to formulate policy or design processes that ensured adequate stakeholder involvement in policy dialogue. Considering that these views were offered by workers in relatively stable governments, the policy capacity of a new Livestock Ministry in South Sudan was a concern. Although it is beyond the scope of this chapter to discuss the wide range of livestock policy options open to the government in South Sudan, a few experiences are worth noting.

In terms of managing policy process, the Livestock Ministry in South Sudan will face a barrage of conflicting ideas on policy. Within the government, it would likely be expected to respond to higher-level directives driven in part by the need for the new government to be seen to be responding to the development needs of people in the south. It will find an incoherent group of aid donors pushing various rehabilitation and development agendas, and in some cases, will need to spend large budgets in short timeframes. Here, South Sudan might learn lessons from other countries in the region that have emerged from long-term conflict. For rebel movements gaining hard-won official recognition and power, there has been a tendency to recreate old government structures for livestock development based on public sector monopolies. A misguided donor response has been to support be wide-scale government construction of veterinary offices, clinics and laboratories, staffed by thousands of employees; large numbers of livestock extension officers could end up joining the government payroll but with limited impact.

Through late 2005 the policy narratives of the southern Sudanese leadership emphasized the general need to strengthen the private sector. For livestock development this means a clear definition of public, private and mixed functions, and policies that avoid competition between government workers and private practitioners. A strong capacity to contract out and monitor a wide range of livestock sector activities will need to evolve. Balancing the political and technical imperatives in the livestock sector will be a challenge for South Sudan – as it is for any other government in any other country.

In addition to government and donor pressures, a third set of policy actors are the Sudanese professionals returning from overseas and taking up new posts in the government. Some have been away for more than 20 years, and while many are very dedicated to helping rebuild their country, the experience of life abroad in relatively wealthy countries may mean they tend to emphasize the need for modernization and new technologies, without fully understanding the past failures of these approaches (see Ashley et al, 1998) and the need for livelihoods-based programmes within the operational context of South Sudan. The NGOs for their part can be expected to call for continued support to assist the poor, commitment to community-based approaches and the need for broad stakeholder involvement in policy development. It is likely that NGOs will continue to attract donor funding and continue to transfer community-level realities to policy makers.

Looking beyond livestock policy, there is the challenge of developing appropriate policies on pastoral and agro-pastoral development. Here the lesson from neighbouring countries is that despite many years of research, projects and good intentions, 'pro-pastoralist' policies are difficult to find. Line ministries in most countries lack appropriate pastoral development policies and pastoral areas remain conflict-prone, underdeveloped and subject to repeated relief interventions dominated by food aid. Pastoral production systems are largely misunderstood by policy makers and negative attitudes towards pastoralists dominate the policy arena. In early 2006 the challenge for South Sudan is to avoid these scenarios, primarily by ensuring that pastoralists and agro-pastoralists enjoy political representation (a right often lacking in other countries), and that livelihoods analysis informs future policy directions.

References

Action Aid (1994) *ActionAid-Somaliland review/evaluation*, October, Action Aid, London.

Almond, M. (1987) *A Paravet Programme in South Sudan*, Pastoral Development Network Paper 24c, Overseas Development Institute, London.

Ashley, S., Holden, S. and Bazeley, P. (1998) *Strategies for Improving DFID's Impact on Poverty Reduction: A Review of Best Practice in the Livestock Sector*, Department for International Development, London.

AU/IBAR (2004) *Institutional and Policy Support to the Livestock Sub-sector in Africa*, Regional overview of a preliminary consultation in the Greater Horn of Africa, African Union/InterAfrican Bureau for Animal Resources, Nairobi.

Blakeway S. (1995) *Evaluation of the UNICEF Operation Lifeline Sudan Southern Sector Livestock Programme*, UNICEF-Operation Lifeline Sudan Southern Sector, Nairobi.

Catley, A. (2000) 'The use of participatory appraisal to assess the impact of community-based animal health services: Experiences from southern Sudan', IXth Symposium of the International Society for Veterinary Epidemiology and Economics, Breckenridge, Colorado (7–11 August).

Catley, A., Leyland, T., Admassu, B., Thomson, G., Otieno, M. and Aklilu, Y. (2005) 'Communities, commodities and crazy ideas: changing livestock policies in Africa', *IDS Bulletin* 36/2: 96–102.

Coalition for International Justice (2006) *Soil and Oil: Dirty Business in Sudan*, Coalition for International Justice, Washington DC.

Deng, F.M. (1987) *Tradition and Modernization: A Challenge for Law Among the Dinka of the Sudan*, Yale University Press, New Haven and London.

Deng, L.B. (1999) 'Famine in the Sudan: causes, preparedness and response. A political, social and economic analysis of the 1998 Bahr el Ghazal famine', IDS Discussion Paper 369, Institute of Development Studies, Brighton.

Deng, L.B. (2002) 'Confronting civil war: a comparative study of household assets management on southern Sudan', IDS Discussion Paper 381, Institute of Development Studies, Brighton.

Donini, A. (1996) 'The policies of mercy: UN coordination in Afghanistan, Mozambique and Rwanda', Occasional Paper No. 22, The Thomas J. Watson Jr. Institute for International Studies, Brown University, Providence.

Fielding, W., Gullick, C., Coutts, P. and Sharp, B. (2000) *An Introduction to the Food Economy Research in Southern Sudan, 1994–2000*, World Food Programme and Save the Children-UK, Nairobi.

Guijt, I. (1998) *Participatory Monitoring and Impact Assessment of Sustainable Agriculture Initiatives*, SARL Discussion Paper No. 1, International Institute for Environment and Development, London.

Harmer, A. and Macrae, J. (2004) Beyond the Continuum: The Changing Role of Aid Policy in Protracted Crises', HPG Report No. 18, Overseas Development Institute, London.

Hopkins, C. (2002) 'Emergency veterinary support to livestock owners in war affected areas of southern Sudan. Field-based PRA training and community participatory evaluation report, Shilluk Kingdom, VSF-Germany, Nairobi.

Hopkins, C. (2003) 'Community participatory evaluation of the OLS livestock programme. A community-based animal health project implemented by VSF Belgium, Bahr el Ghazal, South Sudan', VSF-Belgium, Nairobi.

Hopkins, C. and Short, A. (2001) 'Participatory impact assessment and evaluation of the OLS livestock programme. A community-based animal health project implemented by VSF Suisse, Western Upper Nile Region of South Sudan', VSF-Suisse, Nairobi.

Hopkins, C. and Short, A. (2002) 'Participatory impact assessment and evaluation of the OLS livestock programme. A community-based animal health project implemented by VSF Suisse, Bahr el Ghazal, South Sudan', Nairobi: VSF-Suisse.

Jack, J. M. D. (1961) 'The Sudan'. In G.P. West, Sir F. Ware and the British Veterinary Association (eds) *A History of the Overseas Veterinary Services*, Part One, British Veterinary Association, London.

Jones, B. (2001) *Review of Rinderpest Control in Southern Sudan, 1989–2000*, African Union/Interafrican Bureau for Animal Resources, Nairobi.

Jones, B., Araba, A., Koskei, P. and Letereuwa, S. (2003) 'Experiences with community-based and participatory methods for rinderpest surveillance in parts of southern Sudan'. In K. Sones and A. Catley (eds) *Primary Animal Health Care in the 21st Century: Shaping the Rules, Policies and Institutions*, proceedings of an international conference, Mombasa, 15–18 October 2002, African Union/Interafrican Bureau for Animal Resources, Nairobi.

Keeley, J. and Scoones, I. (2003) *Understanding Environmental Policy Processes: Case Studies from Africa*, Earthscan Publications, London.

Leyland, T. (1996) 'The case for a community-based approach with reference to southern Sudan'. In FAO (ed.) *The World Without Rinderpest*, FAO Animal Production and Health Paper 129, pp. 109–120, FAO, Rome.

Minear, L. (2002) *The Humanitarian Enterprise: Dilemmas and Discoveries*, Kumarian Press, Bloomfield, Connecticut

NSCSE (2004) 'Towards a baseline: best estimates of social indicators for southern Sudan', NSCSE Series Paper 1/2004, New Sudan Centre for Statistics and

Evaluation, Nairobi, www.sudanarchive.net/cgi-bin/sudan?a=d&d=D11d5 [Accessed 8 January 2007].

Okoth, S. (2001) *SC UK Livestock Project 2001 Monitoring and Evaluation Report*, Save the Children-UK, Nairobi.

Tunbridge, L. (2005) *Saving Lives and Livelihoods: Ten Years of Community-Based Animal Healthcare in Southern Sudan*, ITDG Publications, London.

PART II

Case Studies from Somalia

CHAPTER 5
Crisis and food security profile: Somalia

Peter D. Little

Somalia represents one of the most complex and protracted political crises in the world today. The political and institutional environment is so dynamic that information and analysis becomes obsolete very quickly. Writing about it is like trying to focus on a rapidly moving, blurred target: each of the several times that I drafted this introduction, in late 2006, political events forced me to rewrite large sections of it, and the situation continues to evolve quickly even as I write (in early 2007). It is likely to have moved in new directions by the time this chapter is being read, probably challenging large parts of what is written here. During 2006 alone, the country, without a central government since 1991, saw the movement of an exiled Transitional Federal Government (TFG) from Nairobi, Kenya to Baidoa (southern Somalia); the rise and quick collapse of a loose US-backed alliance of militias and warlords called the Alliance for the Restoration of Peace and Anti-Terrorism; the consolidation and control of almost all of southern and central Somalia under the Union of Islamic Courts (UIC); and the invasion by Ethiopian troops and the imposition of the TFG in Mogadishu (in December). It is too soon to forecast whether the TFG will be able to consolidate and pacify the competing factions of Somalia, and to overcome popular support for the UIC and the re-emergence of warlord-based politics. Thus this overview limits itself to the key economic, social and political aspects of Somalia during the period from about 1995 to 2005, before the recent political changes of 2006.

Regional factions and conflicts

After the late President Siad Barre was overthrown in 1991, Somalia was propelled into two years of large-scale conflict and famine and a humanitarian crisis of historical proportions that claimed more than 100,000 lives. It also sparked the exodus of thousands of Somalis, which by 1999 had reached an outflow of more than 1 million people, or the equivalent of about 15 per cent of the country's 1990 population. Many of these ended up in refugee camps in Kenya and eastern Ethiopia, while others sought political asylum in Europe and North America. Since the departure of the UN and US-backed international peacekeeping and humanitarian force in 1994, the country has moved from broader regional conflicts into localized divisions and struggles, often within the same clan group or locality (for background details, see Little, 2003.) Until the most recent events (2006), the conflict was more typical of a 'not peace, not war' situation than a broad-based civil war (Richards and Helander, 2005), but with frequently changing alliances and political actors.

The long period of statelessness has been characterized by many significant political developments, although the consolidation of a central government has not been one of them. Most important has been the establishment of an independent Somaliland in the former British protectorate of northern Somalia, complete with a functioning administration, parliamentary-style government and improved security (Brons, 2001). Less spectacular was the creation of the regional state of Puntland in northeastern Somalia, which, unlike Somaliland, never sought independence, but still established an administration and political institutions.

In the south and central parts of the country, the political changes and rise and fall of different faction heads since 1991 have been too numerous to list here. Mogadishu alone has seen the rise and fall of several powerful warlords and, finally, the consolidation of the city under the UIC in 2006. Regional politics in the Jubba also have changed considerably since the collapse of the state;[1] they hold little resemblance to the political landscape and actors that were dominant in the 1990s. For example, no single group in the area has been able to sustain political dominance for more than a few years at a time nor to extend regional control. In 1999 the Jubba Valley Alliance (JVA), with strong support from Mogadishu-based clan factions, wrested control of Kismayo town from General Hersi 'Morgan' and his militia who controlled Kismayo intermittently from about 1993 up until the JVA ousted him and his militia in 1999. On a few occasions Morgan has unsuccessfully tried to retake Kismayo town but now he has moved out of the area. His Harti clan-backed group, which at one time had support from the powerful Marehan clan, were chief rivals for regional power with the Ogadeen clan (for details on this period, see Little, 2003.) The Ogadeen, especially its Mohamed Zubeyr subclan, was behind the Somali Patriotic Movement (SPM), which helped to oust Barre in 1991 and was a major political entity in the Jubba area up until the late 1990s. When the city fell to the JVA many of the remaining Harti migrated to Puntland while some Kismayo-based Ogadeen businessmen allied with the JVA.

An economy without a state

It is important to recognize that while Somalia has been unable to reconstitute a national political system, it has been successful in re-establishing a national monetary and market system that actually extends well into Somali regions of neighbouring countries (for example, northeastern Kenya and eastern Ethiopia).[2] These have formed an important enabling environment for economic practice in the country and have ameliorated in part what would have been an even greater economic crisis in the region. The re-establishment of monetary and market systems occurred within two years of the state's collapse in 1991. As Marchal (2002) points out, initially there was so much conflict and poor communications between different regions that there were wide discrepancies in monetary exchange rates and considerable variations in local consumer price indexes. However, by early 1993 a national currency market based on the Somali shilling and supplemented by the US dollar had been re-established and halted arbitrage practices; it also restored confidence in the market and currency. This achievement is so important for Somalia because trade is the key economic activity and it can be strongly hampered by an unstable (or absent) currency system. Somali traders involved with the growing cross-border livestock and grain trade with neighbouring countries claim their businesses were aided by the continued convertibility of the Somali shilling (Little, 2003). For poor consumers of rural households an unstable currency also can translate into local food shortages, unstable and high prices and deteriorating terms of trade. All of these developments are especially crippling to low-income consumers and food security generally.

An important reason the country has been able to maintain a monetary system is the dynamic and competitive nature of the informal finance system and the massive imports of remittance incomes. Although already in place prior to 1991, informal money transfer companies (called *hawalidaad* or *hawala*) grew rapidly in the 1990s and 2000s. They are used by individuals, traders and businesses to receive and remit income from the more than 1 million Somalis living outside the country, as well as to transfer money throughout the country, the East African region and the world. In smaller market towns of southern Somalia, these systems operate through a series of small operators using VHF/hand-held radios, but most operators work as agents for large money transfer companies. These informal finance enterprises operate with incredible efficiency, a point that has been noted by many observers (UNDP, 2001; Hansen, 2004), and they annually transfer into the country an estimated $750 million to $1 billion in remittances from the Somali diaspora (Lindley, 2005). What they also do is to help integrate the national monetary system and maintain exchange rate stability within and between different regions and towns. This enormous injection of dollars into the country shores up the Somali shilling, as well as allowing thousands of Somali households to cope with poverty and food insecurity by supplementing their incomes.

This institutional innovation in finance has been developed in tandem with an increasingly sophisticated network of telecommunications, mobile phone network, internet services and other communication technologies. The growth of the telecommunication sector was strongly influenced by the need to facilitate financial transactions in the remittance and trade sectors. In fact, many of the larger money transfer companies also have considerable investments in the telecommunication system since it requires these technologies to operate efficiently. Again, this development has helped to better integrate currency markets in the country and improve market information, as well as commodity markets generally. Thus, a secondary outcome of the country's growth in money transfer companies has been an increase in communication technologies and, consequently, better market information for a range of commodities.

Another macro development that has had strong implications for food security and development in the country is the rapid growth in the commercial trade sector since 1991. Somalia has effectively become an entrepôt economy for large parts of the Horn of Africa, taking advantage of its geography (proximity to the Middle East, especially Dubai) and stateless environment. It imports 'duty-free' rice, sugar, electronics and other goods, which are then re-exported through southern Somalia's large and porous borders into neighbouring Kenya (primarily) and Ethiopia (secondarily) – with few, if any, border taxes. Few Somali experts would argue that overall trade (both import and export) and the private sector generally have not done reasonably well since the collapse of the government. In fact, available data indicate that the value of exports and imports is actually higher now than in 1990 (World Bank, 2005). During the 1990s, livestock exports to the Middle East exceeded the country's pre-1991 levels, although in recent years they have declined following a ban imposed by Saudi Arabia in 2000 due to an outbreak of Rift Valley fever.

This large growth in trade-based activities, including increased import business with Dubai, has helped to integrate market prices and keep supply distribution much better than what would be expected given the amount of conflict that has occurred in southern Somalia. Trucks with consumer good and foods move between the different regions to supply deficit zones even when there are ongoing conflicts. As would be expected, these trade linkages are vulnerable to conflict-induced disruptions and problems and this has had as much impact on local food security as other kinds of shocks (such as drought and floods).

A society in need of security and public institutions

The lack of effective public institutions for maintaining security and guaranteeing and protecting property rights has significantly hampered Somalia's development and has instigated conflict. After 1991 powerful warlord factions moved into the country's valuable riverine areas (Jubba and Shebelle) and stole large amounts of valuable lands and properties from existing owners, often forcing them to become tenant farmers on their own farms. Equally disturbing has been the large-scale expropriation of properties in urban centres where many former residents lost their houses, businesses and land to outsiders. The issue of land and property rights must figure strongly in any post-conflict settlement and development, and some experts have implied that until property and land are restored to their proper owners in the valley, meaningful development cannot take place. As Schlee pointed out at the 2003 Somali National Reconciliation Conference in Kenya, the committee on land and property rights 'was unanimous in its conviction that the prolongation of injustice, that leaving property in the hands of those who have taken it by force, would lay the seed for the next conflict. A lasting peace can only be built on justice, and that is giving property back to those from whom it was taken by force' (2003). And, importantly, this includes large parts of the riverine areas, where continued nutritional and food security problems validate the hypothesis that food emergencies/famines are mainly the result of political processes and land-based injustices.

Public health and education institutions are also sorely required in Somalia. Without a government Somalia has faced its share of public health problems during the past decade. At least three times during the 1990s cholera epidemics killed hundreds of Somalis in the border region, and the El Niño storms of 1997–1998 created a massive malaria and Rift Valley fever outbreak that went uncontrolled for more than one month. As recently as 2006, there was another major outbreak of cholera in southern Somalia. Public health institutions to confront such hazards are absent and the country's expected life expectancy (mid-40s) is among the lowest in the world.

The collapse of the public education system has been particularly damaging for both primary and secondary school-age children. Indeed, the term 'lost generation(s)' accurately captures the current predicament of students in Somalia, since even the most optimistic statistics indicate 'only 20 per cent of school-age children (ages 5–14) are enrolled in school, and are heavily concentrated in the early grades' (UNDP, 1998). The participation rates are even lower in rural areas, especially in the pastoral zones. The predicament for secondary and post-secondary education may be even more desperate. Attendance at secondary and post-secondary schools is only a small fraction (probably less than 25 per cent) of what it was in the 1980s. Some estimates claim that fewer than 2,000 children attend secondary schools in Somalia (excluding Somaliland and Puntland), and most of these live in Mogadishu, where the majority of high schools are located (UNDP, 1998). The lack of education and other opportunities partially explains why warlords have been able to recruit extensively from among the youth. Often armed with Kalashnikovs (AK-47s) and attached to heavily armed vehicles called 'technicals', the Somali youth have lost an opportunity to acquire the skills needed to rebuild institutions and industries and to compete regionally. Without access to education their ability to diversify into higher-return activities – as some of their pastoral neighbours such as the Maasai of Kenya have done to supplement mobile pastoralism – will be limited.

Humanitarian and development responses

Food security-related programmes in Somalia have focused mainly on 'keeping people alive' in the short term through food aid, cash relief in a few cases, emergency medical and water provision, seed distribution, irrigation rehabilitation and other short-term efforts. Most externally funded projects have been implemented with funding horizons of half a year to two years, although some NGOs have been in certain locations and in certain sectors for several years (for example the ICRC and World Vision). The use of food aid has dominated emergency assistance and, in the case of the US, the food aid has been sourced outside eastern Africa. There are numerous NGOs, both local and international, that have been involved in emergency responses. These efforts are necessary to avoid catastrophic humanitarian loses but generally are not the basis for longer-term recovery in southern Somalia.

The regional politics and security situation in Somalia have been so fluid that most donors and NGOs work directly with local community and village bodies and district councils rather than with so-called regional authorities such as the former JVA. In the absence of counterpart national and regional institutions in Somalia, the donors and NGOs have established the Nairobi-based Somalia Aid Coordination Board (SACB) for coordination, information sharing and a level of priority-setting and planning around key issues such as food security. SACB operates around sets of working groups, including groups that focus on food aid and rural development. These bodies deal with general humanitarian and development strategies, share information and data and plan for the coming year. Because most donors and NGOs initially assumed that Somalia would have a central government within a reasonable period of time, they often did not deal with long-term plans and strategies and assumed that viable national counterpart institutions would make SACB unnecessary.

The European Commission (EC), which is the major donor agency in Somalia, has increasingly funded programmes with terms of two years or more. In 2002 it developed a five-year (2002–2007) strategy for Somalia entitled 'Strategy for the Implementation of Special Aid to Somalia' (SISAS), which informs its different programmes in the country (EC, 2001; 2002). In addition to humanitarian assistance and conflict reduction, SISAS has a strong focus on food security and rural development. In 2006 the EC was to have provided 10.5 million euros to six different rural/agricultural development projects, including support to market development and rainfed agriculture in south and central Somalia (EC, 2006). In 2006 it proposed to finance a 'Livelihoods Support Programme in South Somalia' that 'will aim at assisting households' livelihood recovery, stopping further depletion of assets, improving purchasing power and increasing resilience to future shocks' (EC, 2006). While the EC has been strongly limited in what it can do in the Jubba area because of political conflict, it does have a strategy in place for moving from humanitarian assistance to development assistance (EC, 2001).

In the Lower Shabelli Region of southern Somalia the EC-funded Shabelle Agricultural Rehabilitation Programme (SHARP) is an example of the agency's willingness to move beyond humanitarian assistance. SHARP has an approach that relies on 'the use of participatory methods, and efforts that work toward capacity-building, partnerships with local communities, and meaningful participation by beneficiaries' (Longley et al, 2005). SHARP is one of the few longer-term development efforts in the south and while it is too early in its implementation to effectively assess its impacts there have been some concerns raised about its potential political impacts. As Longley et al (2005) caution about the SHARP programme:

> It cannot be assumed simply because the EC has been able to implement developmental interventions that this will necessarily have positive impacts not only in terms of food security but also in terms of political developments in the long term. ... It is apparent that many interventions within the food security sector are implemented with almost complete disregard for the ongoing conflict.

The stakes for a programme like SHARP are especially high because struggles over irrigable land already have resulted in considerable conflict and these could easily be further aggravated by a poorly conceived development programme. Thus, while the EC is willing to take far greater risks in its assistance programmes in Somalia than other donor agencies, the activities can have both positive and (as discussed here) negative outcomes.

Timeline of conflict in Somalia

The following chronology represents the key events in Somalia from the collapse of the government in the early 1990s through to December 2006 (based in part on UNDP, 2001).

January 1991 Collapse of Siyad Barre regime; President Barre flees Mogadishu.

May 1991–December 1992 Famine conditions develop and widespread war takes hold between different warlords and their militias.

May 1991 Somaliland (northwestern Somalia) declares independence.

April 1992 United Nations Operation in Somalia (UNISOM) created.

December 1992 US Marines land in Mogadishu and US-led 'Operation Restore Hope' launched in collaboration with UN campaign.

October 1993 Eighteen US military killed in Mogadishu in clash with warlords and militia. US announces plans to withdraw from Somalia.

March 1994 US withdraws from Somalia.

January 1995 Former President Barre dies in exile in Nigeria.

March 1995 UNISOM forces withdraw from Somalia without establishing a central government.

June 1995 General Mohamed Farah Aideed declares a government but it is not recognized internationally and fails to unite the country.

August 1996 Aideed dies in gun battle. His son Hussein Aideed takes over.

Janaury–February 1998 Floods and outbreak of Rift Valley fever in southern Somalia leads to ban on livestock exports to Saudi Arabia.

May–August 2000 The Somali National Peace Conference is held in Arta, Djibouti and the Transitional National Government (TNG) is established, with Abdiqasim Salad Hassan elected as president. Government moves to Mogadishu in October 2000.

September 2000 A second ban imposed on Somali livestock exports after several Saudis die after consuming imported meat.

October 2002 After strong dissatisfaction with the TNG and its leaders, the 14th Somali peace/reconciliation conference is convened in Kenya.

October 2004 Somali delegation at reconciliation conference in Kenya elects Abdullahi Yusuf as president and establishes TFG (initially based in neighbouring Kenya).

September 2004–December 2005 Severe drought and humanitarian crisis in southern Somalia.

August 2005 TFG moves to Somalia and establishes temporary capital at Baidoa. Outside of Baidoa TFG control over country (including Mogadishu) is very limited.

February 2006 Fighting between UIC and Mogadishu-based warlord erupts and lasts until May. It is the worst fighting and most casualties in the city since the early 1990s.

June–July 2006 UIC defeats warlords and takes control of Mogadishu and other parts of the south.

August–September 2006 UIC re-opens Mogadishu seaport and airport; relative peace is established in area. Meetings between UIC and TFG officials are held in Sudan but fail to resolve differences between the two.

October–November 2006 Open war between TFG and UIC seems imminent and thousands of Somalis flee to neighbouring country. Drought and humanitarian crisis continue in parts of southern Somalia

December 2006 Ethiopian and TFG forces move on Mogadishu. UIC militia is overrun and flees to the south.

References

Brons, M. H. (2001) *Society, Security, Sovereignty, and the State: Somalia, from Statelessness to Statelessness*, International Books, Utrecht, Netherlands.

EC (European Commisson) (2001) *Linking Relief, Rehabilitation and Development – An Assessment*, EC, Brussels.

EC (2002) *Strategy for the Implementation of Special Aid to Somalia (SISAS)*, EC, Brussels, www.delken.ec.europa.eu/en/information.asp?MenuID=3&SubMenuID=13#bri [accessed 16 January 2007].

EC (2006) *Livelihood Support Programme in South Somalia*, EC, Brussels.

Hansen, P. (2004) *Migrant Remittances as a Development Tool: The Case of Somaliland*, Working Paper Series No. 3, Danish Institute for International Studies, Copenhagen.

Lindley, A. (2005) *Somalia Country Study*, ESRC Centre for Migration, Policy and Society, Oxford, UK.

Little, P. D. (2003) *Somalia: Economy Without State*, James Currey Publishers, Oxford, UK and Indiana University Press, Bloomington, IN.

Longley, C., Hemrich, G. and Haan, N. (2005) *Reconfiguring Aid: Food Security Programming and Policy in Somalia*, FAO, Rome (mimeo).

Marchal, R. (2002) *A Survey of Mogadishu's Economy*, Somali Unit, European Commission, Nairobi.

Menkhaus, K. (1999) *Studies on Governance in Lower Jubba Region*, United Nations Development Office for Somalia (UNDOS), Nairobi.

Menkhaus, K. (2003) *Somalia: A Situation Analysis and Trend Assessment*, United Nations High Commissioner for Refugees (UNHCR), Geneva.

Richards, P. and Helander, B. (2005) *No Peace, No War: An Anthropology of Contemporary Armed Conflicts*, James Currey Publishers, Oxford, UK.

Schlee, G. (2003) 'Consultancy report to the Somali National Reconciliation Conference in support of Committee 3 – Land and property rights', IGAD, Djibouti.

UNDP (United Nations Development Programme) (1998) *Human Development Report: Somalia*, UNDP, Nairobi.

UNDP (2001) *Human Development Report: Somalia*, UNDP, Nairobi.

World Bank (2005) *Somalia: From Resilience Towards Recovery and Development*, Report No. 34356-SO, World Bank, Washington DC.

CHAPTER 6

Livelihoods, assets and food security in a protracted political crisis: The case of the Jubba Region, southern Somalia

Peter D. Little

Abstract

This chapter addresses livelihoods, poverty and food insecurity in the Jubba region of southern Somalia, highlighting the key roles that market access, diversification and assets have played in allowing certain livelihoods to adjust to a highly unstable environment while others have become more vulnerable. It suggests that important social, demographic and economic changes were already in place prior to the protracted crisis, but that they have been sharply distorted by the current situation. It also considers what might be entailed in a longer-term development approach for this area and how this relates to information and institutional needs.

Introduction

The Jubba area[1] of southern Somalia (see Figure 6.1) has been among the most unstable regions in one of the world's most volatile countries. Since the collapse of Somalia's government in 1991, the region has experienced a war-induced famine with more than 60,000 human deaths (1991–1992) (de Waal, 1997; Prendergast 1997); periodic outbreaks of cholera and other public health threats such as malaria; endemic political insecurity that has isolated the region's major city (Kismayo); and at least four major droughts and five flood episodes. The most recent drought (2004–2005) and floods (2006) have been among the Jubba's worst natural disasters in the past 50 years.

The effects on local populations of these frequent shocks, however, have been uneven. Parts of the region, especially the pastoral areas in the east, have done relatively well, while other locations have experienced massive destruction, malnutrition and food insecurity (FSAU/UNICEF, 2004). Especially the Jubba Valley, which is inhabited by a mix of groups comprised mainly of politically marginalized Bantu groups (often called *Gosho* people), has been negatively impacted and has seen the exodus of tens of thousands of its residents. Although the humanitarian situation has not been as severe in Kismayo town, it also has experienced widespread economic and social dislocations and the destruction of much of its infrastructure. In short, the Jubba area represents a social and political microcosm of Somalia itself, with its oppressed and chronically food-insecure minority groups, an embattled urban population and a mobile pastoral population that is doing moderately well under very difficult circumstances.

This chapter addresses livelihoods, poverty and food insecurity in the Jubba region of southern Somalia with a focus on the period from 1995 to 2005. It highlights the key roles that market access, diversification and assets have played in allowing certain livelihoods to adjust to a highly unstable environment when others have become more vulnerable. It suggests that while important social, demographic and economic changes were already in place prior to the protracted crisis, they have been strongly distorted by the current situation. These processes include increased vulnerability to natural disasters (for example, drought and floods), diversification of livelihood systems, heightened dependence on markets, inequality and poverty, and decreased per capita holdings of key assets (especially livestock). The chapter proposes an asset-based model for examining the relationship between household welfare and conditions of poverty and food insecurity. By utilizing an asset-based approach, it is possible to depict how Jubba households move in and out of different states of poverty and food vulnerability in response to various kinds of shocks, including conflicts, droughts and market disruptions.

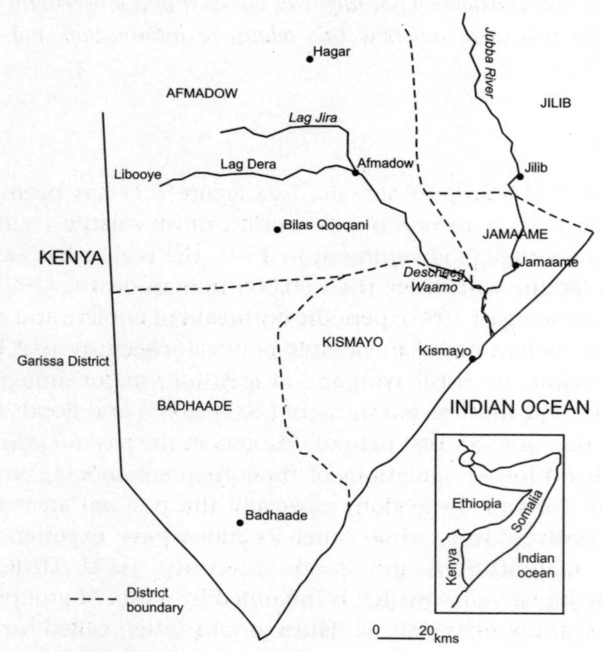

Figure 6.1 Jubba area, southern Somalia
Source: Little (2003)

The chapter begins with an overview of the Jubba region and its economic and social potential and problems. It then turns to an assessment of the area's livelihood systems, with an emphasis on the key roles played by household assets and trade. By highlighting the dynamic nature of poverty and the need to distinguish chronic from transitory poverty and food insecurity, the discussion goes beyond existing works on livelihood systems in Somalia. It addresses some of the longer-term causes of poverty and food insecurity in the area, as well as some of the immediate factors that result in localized emergencies. The final part of the chapter concludes with assessments of institutional responses to the Somali situation and different approaches to food security and development in the Jubba area. It also emphasizes the context-specific characteristics of the region that need to be part of any policy-based framework for moving from relief/humanitarian to longer-term development programmes. The dynamic nature of political and economic events is discussed to highlight the need for flexibility and careful monitoring of any policy-based initiatives.

It should be stated at the outset that the Jubba area, like many parts of Somalia, suffers from a chronic scarcity of reliable information. Perhaps only the northern (Somaliland) and northeastern (Puntland) regions of the country have even a bare minimum of economic and social data that could allow for some modicum of development planning (UNDP, 2001; World Bank, 2005). What data have been generated in the south have come from short-term field investigations and project appraisals, which often become quickly dated. Nonetheless, available secondary data and published and 'grey' literature were heavily consulted in writing this chapter. The author has also relied on his previous research in the Jubba area in the late 1980s and in the border areas of southern Somalia/northeastern Kenya during 1996 and from 1998 to 2002.

Background to the Jubba region

The Jubba region is an important economic zone for several reasons. Most important are the facts that it encompasses a major deep sea port at Kismayo town; the country's only true perennial river (Jubba);[2] large expanses of irrigable and rainfed agricultural lands; a long border with East Africa's largest economy (Kenya); and Somalia's largest cattle herds and a sizeable proportion of its camels. Conflicts between different clans and militia groups often centre on the region's key economic resources. For example, fighting over the control of Kismayo town and its port facilities has been almost endemic since 1991 with widespread human suffering and population displacements (UNOSOM, 1994; Narbeth and McLean, 2004). In late 2005 a WFP ship with food aid was hijacked off the Kismayo coast and a national employee of the UN was killed in Kismayo. The latter tragedy halted plans for the re-establishment of UN offices and programmes in the Kismayo area.

The Jubba Valley, in turn, has experienced widespread armed struggles over its valuable lands that have devastated sedentary farmers and minority groups, such as the *Gosho* (Menkhaus and Craven, 1996; FSAU/UNICEF, 2004). With the collapse of the state in 1991, Mogadishu-based factions came south and pillaged the Jubba area, grabbing up important chunks of irrigable land and maintaining an important presence that still intimidates local farmers. They also took over key properties and facilities in the Kismayo port area.

In comparison with central and northern Somalia, the Jubba area has a complex ethnic complexion. During the past 150 years, it has served as a frontier zone for the expansion of major clans, such as the Ogadeen and Harti, as well as smaller groups. The social composition of clans and subclans is especially complex east of the Jubba River, where one finds more than 10 different clans, subclans and Bantu communities claiming historical rights to different areas. The area's history continues to complicate current politics. For example, the Marehan, who had strong government support from the late President Barre, staked claims to large tracts of irrigable land in the middle and lower Jubba Valley and numerous businesses in Kismayo town. By the time the state collapsed, the Marehan were strongly embedded in the Kismayo business sector, including the activities of the port, and also controlled lucrative lands along the Jubba River.

Not surprisingly, regional politics in the Jubba have changed considerably since the collapse of the state.[3] Many earlier faction leaders (warlords) have disappeared, but the key political spoils and targets remain the same (Little, 2003; Menkhaus, 2003). First and foremost is Kismayo town and its port, which provides steady monthly revenues for whoever controls it. It is estimated that the port earns up to $100,000 monthly on 'taxes', security fees and other charges for the local political faction that controls it. Until 2006 the Jubba Valley Alliance, a fragile union of Haber Gedir (Hawiye clan) militia and businessmen from the Mogadishu area and the late President Barre's Marehan clan members, ruled the port. Its revenues were the JVA's major source of income (for additional details on the JVA, see Menkhaus, 2003).

The second key target of political conflict has been the Jubba Valley and its rich, well-watered agricultural lands. Once an important agricultural surplus area and source of export revenues (mainly from bananas) for southern Somalia, it is now among the most food-insecure and impoverished areas in the country. When the state collapsed in 1991 the riverine area was, so to speak, 'a low-lying fruit' on Somalia's fractured landscape. Militia groups coming down from the Shebelli areas and Mogadishu and nearby pastoralists of the Lower Jubba grabbed up large chunks of the valley's valuable farmland and destroyed much of its infrastructure. It will take massive public investment to repair and bring back much of the irrigable land in the area and to control the periodic floods which, with the loss of embankments and control canals, now occur almost annually. It will take even more effort to resolve the land disputes and conflicts in the valley that continue to colour the area. Reasons for the latter challenge include the fact that: (1) land was illegally and sometimes forcefully taken both during the 1980s, with bogus land titles granted to 'owners', and after the state collapsed in 1991, which complicates the question of who are the rightful occupants/owners pre-1991; (2) many of the rightful owners of riverine land were forced to flee the area, including several thousand *Gosho* people who resettled in the US; and (3) remaining minority residents who have the strongest claims to valley lands have little voice in current political dialogues and alliances in Somalia.

Livelihoods

Types of livelihoods

Using the livelihood classification system of the Food Security Analysis Unit (FSAU)-Somalia, the Jubba area contains parts of the following livelihood zones (FSAU, 2004):

- Jubba pump irrigated (JPI) commercial farming (tobacco, onions and maize);
- southern Jubba riverine (SJR) (maize, sesame, fruits and vegetables);
- lower and middle Jubba agro-pastoral (LMAP) (maize and cattle);
- southern inland pastoral (SIP) (camels, goats and sheep);
- south-east pastoral (SEP) (cattle, sheep and goats).

As noted earlier, the riverine area still experiences some of the country's most serious human rights abuses and conflicts, nutritional problems and chronic poverty. In the Jubba Valley there is a narrow band of commercial and subsistence irrigation along the river that is represented by JPI and SJR in the above classification system. However, this was not always the case. Prior to the collapse of the state there were flood control works along the river that stabilized water flow, commercial banana estates and three large state-sponsored irrigation schemes. The latter projects were looted during 1991–1992 and it was said that some of the stolen infrastructure was exported to the Middle East as scrap metal. The commercial banana estates, in turn, were taken over by different factions after 1992 but they soon collapsed. The banana industry, which was revived for just a short period during 1995–1997, has left many households in the lower riverine area without an important source of employment. In the past 10 years, some fruit and sesame production has replaced banana production in the riverine zones and provided employment for low-income farmers.

The pastoral and agropastoral production systems (LMAP, SIP and SEP) in the Jubba region have proved more resilient than the crop-based systems described above. They rely heavily on the production of cattle, goats and camels. Important differences in these systems exist based on the distribution of different livestock species and local water and vegetation conditions. Cattle pastoralism is concentrated in areas of expansive grasslands to the west and southwest of the region, while camel and goat production are found in the interior where browse species and acacia trees prevail. In the northwestern parts of the region (for example, western Afmadow and Hagar districts), where browse species dominate, herders also keep large numbers of camels. The reasons for pastoral resilience in the region include:

- *Mobility*. Despite periodic constraints caused by insecurity, livestock-based systems have remained relatively mobile. Mobility has allowed livestock owners to cope with local variations (temporal and spatial) in rainfall and pasture condition and occurrences of conflict.
- *Cross-border trade*. The area's proximity to Kenyan (primarily) and Ethiopian markets have allowed local households to earn market income from this trade and/or to work as trekkers, transporters, loaders and off-loaders and security persons in the commerce. Since 2000, the Jubba area has supplied an estimated 70,000 animals annually to the Kenyan meat market (Little, 2005).
- *Remittances*. As discussed later in the chapter, regional remittances are important for Jubba families. These have helped pastoral households to cope with income shortfalls and, in some cases, to invest in small businesses and other non-farm activities.

Increased livelihood diversification

Like households in other risk-prone environments, Jubba families diversify their income sources in order to spread risk and cope with drought and general insecurity (Little, 1992; Little et al, 2001). Those households that have been able to combine successfully different income-earning activities (for example, trading and rainfed farming) with livestock production have proved more resilient than less diversified households. Some of this diversification has been a 'natural' response to increased land pressure and market opportunities and was occurring prior to the government's collapse, but other changes stem from recent factors (often political in nature). For example, increased cultivation among pastoral and agropastoral communities in the Jubba has been a 'normal' response to decreased per capita livestock holdings and volatile grain prices, a pattern that has been observed in neighbouring countries as well (McPeak and Little, 2004). The steep decline in waged employment as a diversification strategy, by contrast, has been an 'abnormal' condition caused by conflict and the collapse of large-scale agricultural enterprises. Prior to 1991 it was very common for poor households in the lower Jubba to pursue casual labour opportunities on commercial banana estates or in Kismayo town.

Although casual labour remains an important source of income for poor households, wage-earning opportunities have declined greatly in recent years. As one report on Jilib notes, 'With the decline in large-scale irrigated agriculture, the closure of many shops and businesses in small towns, and generally fewer casual labour opportunities, current households are highly dependent on rainfed and flood irrigation (*deschek* agriculture)'[4] (FSAU/UNICEF, 2004). Thus, the capacity for poor farmers to manage risk by diversifying into waged employment has diminished today and reduces their ability to escape from poverty and food insecurity.

Reduced job opportunities have been compensated partially by increased remittances. These transfers have helped some households cope with the loss of normal streams of income and, in a few cases, to avoid chronic poverty. A recent national survey suggests that remittances account for 22.5 per cent of household income (UNDP/World Bank, 2003). While international remittances from Europe and North America are generally low in Jubba's rural areas, transfers from neighbouring countries are high. Indeed, the largest number of registered (and unregistered) Somali refugees worldwide is in neighbouring Kenya and, as noted earlier, there is a booming cross-border trade between Somalia and Kenya that strongly involves the Jubba area. It is not uncommon for a rural Jubba family to have a member working as a trader or labourer in Kenya. These migrants frequently remit income to family members and other relatives in southern Somalia. The Eastleigh area of Nairobi, which is a major business centre, is dominated by Somali merchants, some of whom are from the Jubba area. They often remit income and goods back to relatives in Somalia (Campbell, 2005).

Other types of income diversification strategies in the region include charcoal production. This practice is encouraged by large businessmen who buy the charcoal to export to the Middle East, in some cases hiring bands of labourers to clear cut large tracts of remaining acacia forests. Near the Kismayo port one can observe massive stock piles of bagged charcoal awaiting export to the Middle East (author interview data, June 2006). The commodity has come to be called 'black gold', since it can be sold in Saudi Arabia for up to four times more than on the local market. The activity has had the most serious environmental effects in the Jubba Valley and the coastal plains south of Kismayo town, where hundreds of hectares of forest lands have been cleared (Menkhaus, 2003).

Herder cultivation

Where agriculture is feasible, herder households increasingly pursue cultivation, a pattern that was just beginning to take place in the pre-1991 period. Because of the relatively high rainfall in the Jubba region, which is as much as 700 millimetres (mm) per year in the valley, and the presence of the Jubba River, cultivation is a viable option in several parts of the region. It is estimated that poor agropastoral households in 'normal' years can earn up to 50 per cent of their annual cash income from crop sales, while middle-wealth agropastoralists can attain 30–45 per cent of total income from the activity (FSAU, undated). These figures greatly exceed the importance of cultivation during pre-1991 years, when it accounted for less than 3 per cent of household income (Evans et al, 1988).

The first and most important reason why herders became more interested in agriculture is that they own fewer animals than in the pre-1991 period. Human population growth, increased inequities of livestock ownership and the numerous shocks (for example, droughts and war) of the years since 1991 have left many households with too few livestock to rely solely on pastoralism. By growing their own cereals, households are able to hold on to their remaining animal assets and are not required to sell them to purchase food. This tactic assists herd recovery and accumulation after a drought.

A second factor that motivates interest in farming relates to the generally poor terms of trade for livestock producers during the past several years (discussed in more detail later in the chapter). Grain prices are high and unstable in most parts of the Jubba region, while at the same time livestock prices have stagnated (Little, 2005). Increased cross-border cattle trade with Kenya during the past decade helped to ameliorate market problems, but unlike the past it represents the only viable market option and, thus, herders have little negotiating power with traders and receive relatively poor prices (Little, 2005). In the Kenya border trade herders receive the equivalent of about 40 to 45 per cent of final sales prices in Nairobi or Mombasa, Kenya. This lack of viable market options, a direct result of the protracted political instability in the region, has kept livestock prices relatively low in the area, especially when compared to grain and other food (especially imported food) prices. Under these circumstances pastoral and agropastoral households attempt to produce their own grain in order to avoid transactions in an unfavourable market.

Incidences of shocks and their effects

Even in the years since 2000, the Jubba area has experienced an almost endless litany of 'shocks'. Droughts occurred in 2000 and 2004–2005. Major floods, in turn, occurred in at least three of those five years, and many riverine communities struggled to cope with the aftermath of the floods caused by the 1997–1998 El Niño event. In 2006 there was a major flood in the region that created havoc for tens of thousands of residents.

Shocks that affect markets in the area particularly affect how households in different livelihood zones cope with food insecurity and poverty. Similar to other food deficit areas, the poor are more dependent than others on market purchases because their capacity for self-provision through their own farms and herds is low. The poorest households are forced to sell off grains almost immediately after the harvest to pay off debts from the previous year and to purchase household necessities. Jubba households generally rely on the market for about 40 per cent of their food needs during most of the year and up to 70 per cent during the long dry season (*jilaal*) (Little, 1989; also FSAU, undated).

In the late 1980s the Jubba region had a fairly well-functioning regional grain market. Maize and sorghum from surplus areas were sold through a network of private traders and retailers to needy consumers in deficit areas. Price differences between different markets at the time were what would be expected, once transport and other market costs were taken into account (Evans et al, 1988). Intra- and inter-regional grain flows were considerable during those years, including trade with other regions of southern Somalia.

The marketing pattern described above contrasts sharply with the current situation (2006). There are strong distortions due to the numerous roadblocks and militia-enforced transit fees. Because key market and production areas are cut off from each other for long periods of time, there are wide price variations for the main cereal (maize) at the four key regional markets. If one looks only at the price differences for maize between Jamaame (Jubba Valley) and nearby Kismayo town from 2002 to 2004, they vary between 92 to 215 per cent, which is more than five times the price variation in 1987–1988. Even in the same marketplace there can be monthly price spikes for certain commodities as high as 70 to 80 per cent. The combination of market instability and price distortions especially punishes the poorest households who can least afford high prices.

Table 6.1 shows the incidence of different shocks in the Jubba area, including market-related shocks, during 2001 to 2005. They include economic ('trade route closed, currency devaluation/instability'), natural ('drought, flood') and political shocks ('major conflict, border closure'). As the table shows, incidences of political/conflict shocks have been the most common, followed by floods and the closure of trade routes. The latter usually happens because of a conflict between different factions or because the costs of roadblocks and 'taxes' become so high that the trade route is temporarily closed due to loss of profitability. Such actions can result in monthly price increases of 50 per cent or more for basic food staples, which has a devastating effect on the poor who are strong net-purchasers of food.

The occurrence of periodic droughts, illness, conflict and other serious events tends to wipe out asset gains that poor households attain and at the same time increase the number of poor in the area. Because droughts and floods occur frequently in the area, the poor face a situation whereby once they begin to rebuild their assets, the next disaster wipes out the gains and recovery must start over again. During these years, the poorest households have few material assets to sell off and may seek casual labour opportunities, lend off children to better-off households, reduce consumption and/or collect wild foods to cope. Each successive shock increases the pool of poor and vulnerable households in the area. As will be shown in the next section, the cumulative effect has been a steep erosion of household and community assets and an increasingly small number of households who can cope effectively with the volatile environment.

Table 6.1 Monthly incidence of 'shocks,' Jubba Area (Lower and Middle Jubba regions), 2001–2005

Shock	*Number of monthly incidences per year*					*TOTAL*
	2001	*2002*	*2003*	*2004*	*2005*	
Major conflict	3	6	3	3	1	16
Flood	2	2	3	1	3	11
Trade route closed	1	2	1	3	0	7
Drought	0	2	1	1	2	6
Border closure	2	0	2	1	0	5
Currency deevaluation/instability	2	0	1	1	0	4

Source: Based on a review of FEW-NET's monthly bulletins, 2001–2005

The key role of assets

To better understand the food security crisis in the Jubba area, it is important to assess the ways in which key economic assets affect households' abilities to cope and recover from these various shocks. It is important to distinguish chronic from transitory food insecurity, the latter being temporary and shock-driven and the former more permanent. Chronic food insecurity is symptomatic of larger structural problems and extreme conditions of poverty (Barrett and Maxwell, 2005). It is also important to address how poverty dynamics – as defined by ownership of key assets (wealth) – affects these different types of food insecurity and vulnerability and the capacity of households to cope and recover from shocks.

Like food security itself, poverty can be distinguished according to transitory versus chronic or persistent states of poverty. As Little et al (2006b) point out:

> Transitory poverty is associated with movements into and out of income poverty, while chronic poverty reflects persistent deprivation. The former type usually results from a drought or other disaster that knocks a household into poverty for up to a few years. After the shock ends and recovery ensues the household rebuilds its assets and moves back out of poverty. In the case of chronic poverty, however, poverty persists in shock and non-shock years as households control too few assets and are insufficiently productive in using those assets to allow them to escape from poverty without external assistance.

The model in Figure 6.2 shows the relationship among asset holdings, poverty and food insecurity thresholds, and recovery before and after a hypothetical shock. (This model has been applied empirically in Ethiopia; see Little et al, 2002; Carter et al, 2004; Little et al, 2006a).[5] In the model, asset endowments along the 'y' axis largely determine a household's or individual's future capacity to earn income and withstand shocks (i.e. maintain *resilience*). Any type of asset or asset bundle (for example, livestock, land and/or labour) can be plugged into the model. Building up household and community assets to the point that humanitarian crises do not ensue every time a shock occurs – which in the model is a hypothetical level of 3.5 units (the level of animal/asset ownership for many livestock-based livelihood systems, see discussion below) – is the key to sustained development (Longley et al, 2005).

Figure 6.2 shows that the non-poor are able to avoid poverty and hunger by selling off (or consuming) their assets and pursuing other strategies as a shock progresses. The poorest and most food-insecure, in turn, are stuck at very low levels of equilibrium represented by the dotted line at the bottom of the graph. As used in the model, vulnerability refers to a poor household's likelihood (probability) of moving into poverty and food insecurity (hunger) as a result of exposure to a hazardous event (shock). In the words of Hoddinott and Quisumbing (2003), 'vulnerability is the likelihood that at a given time

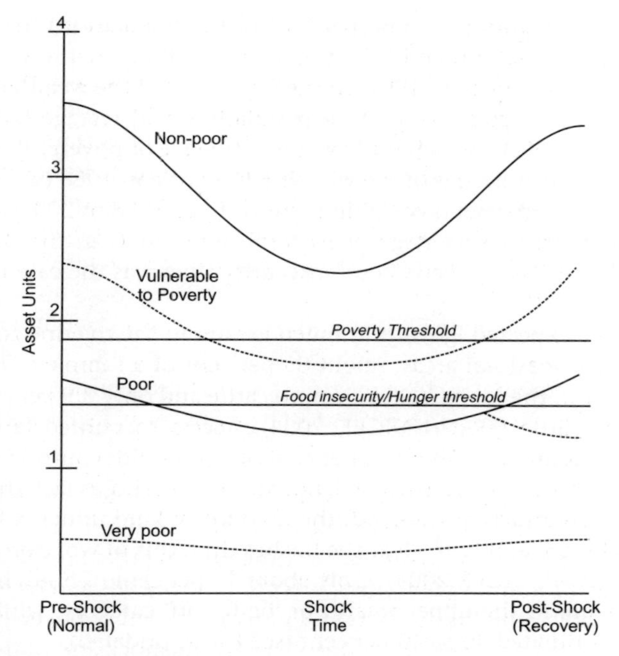

Figure 6.2 Asset shocks and recovery
Source: Author's own research

in the future, an individual will have a level of welfare below some norm or benchmark' (2003). *Resilience,* by contrast, is the ability of a household to withstand and recover from hardships, and return to pre-existing asset and welfare levels within a relatively short period. In the model a household is considered to be persistently or chronically poor if it has been continuously poor for five or more years (see Chronic Poverty Research Centre, 2004).

For pastoral/livestock-based systems the non-poor tend to have animal assets of about 3.5 to 4 tropical livestock units (TLU)[6] per capita but this will vary depending on what other assets (for example, land) are owned by the household (Lybbert et al, 2004; Little et al, 2006). If we compare current livestock assets and wealth categories as defined by the FSAU livelihood system (see FSAU, undated) with the pre-1991 situation several interesting trends can be observed.[7] Prior to the government's collapse, Jubba herders had average livestock assets of about 8.5 TLU per capita (Little, 2003). At the time the wealthiest 25 per cent of households controlled 60 per cent of livestock wealth, while the poorest quartile had rights to only 5 per cent of total herds and had per capita holdings of less than 1.0 TLU. More than 65 per cent of households in the area had the minimum resilience threshold of about 4 TLU per capita.

By contrast, the poorest 27 to 34 per cent of the population in 2004 had average per capita asset holdings of about 1.0 TLU (FSAU, undated). In the middle wealth category (50 to 60 per cent of the population) herd holdings ranged between about 1.6 and 2.7 TLU, a level well below the wealth of the middle two quartiles prior to 1991. At the upper end of the wealth continuum the 'better-off' (15 to 20 per cent of the population) had average asset holdings of about 7.0 to 8.5 TLU, which is above the resilience or poverty threshold but well below average holdings of the wealthiest quartile in 1988 (see Figure 6.3). Thus, the changes in asset ownership patterns from 1988 to 2004 suggest that a considerably larger percentage of pastoral households in the Jubba are in chronic and transitory poverty/food insecurity than was the case prior to the state's collapse.

As would be expected, poverty is much greater in the riverine zone than in the pastoral/agropastoral areas. About 89 per cent of a sample of households in Jilib District in the Jubba Valley own no cattle and only 2.9 per cent possess four or more cattle (FSAU/UNICEF, 2004). Access to rainfed land is not a constraint in the area and about 62 per cent of households control four or more hectares, but very few own irrigable land. Many households lack the resources to fully cultivate much of the land: the amount of land under cultivation 'is determined by the ability to clear the bushes (in terms of workforce and tools required)' (FSAU/UNICEF, 2004). Only about 15 per cent of households in the Jubba Valley are in the upper wealth or 'better-off' category, while the poor make up an estimated 30 to 40 per cent (see FSAU, undated).

In sum, the relationship among asset ownership (wealth and poverty), livelihood strategies and food security is never static, especially in the Jubba area. A livelihood classification system needs to account for the fact that the proportion of a population in each wealth/asset category will vary depending upon where the location is along the shock/recovery ('bust/boom') cycle. A one-shot classification system, especially if it is based on data in an immediate post-drought period, may be misleading and fail to capture the amount of movement between different categories of wealth and food security/ insecurity.

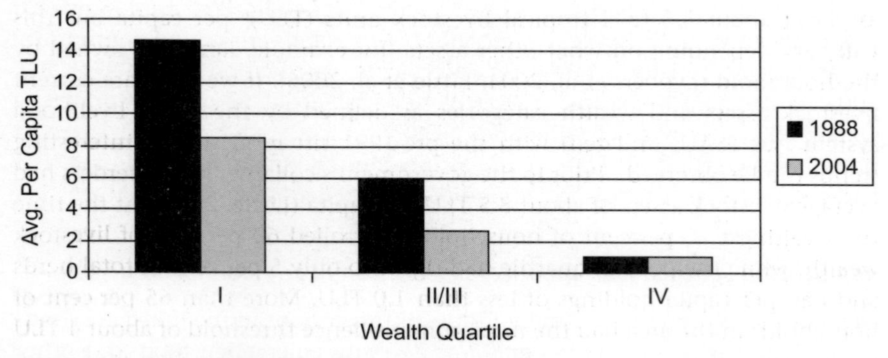

Figure 6.3 Livestock holdings in Afmadow/Kismayo districts, Somalia, 1988 and 2004
Source: Author's own data (1988) and analysis of FSAU data (FSAU, undated); see also Note 7.

Humanitarian and development responses

There have been more than 75 local and international NGOs involved in humanitarian and development work in the region between 1991 and 2005. Those international NGOs that have been in the Jubba area for a significant period of time (for example, World Vision, Terra Nuova and ICRC) have forged important working relationships with local groups. In some cases these partnerships have been sustained despite the short time horizon of most projects. Without a regional or national administration, they mainly work with community and district groups as local partners. Most NGO efforts, including ICRC, tend to have project horizons of two years or less.

In some cases the approach promoted has been of a humanitarian/relief nature and has shown limited understanding of the context. There are numerous examples of well-intentioned NGOs providing emergency services and assistance in southern Somalia on the assumption that functioning markets were not present. For example, in the 1990s one international NGO provided free veterinarian services and medicines to herders when an informal market for these services had been operating since at least the 1980s (see Little, 1989; Catley, 1999); food aid was provided by NGOs in areas where they mistakenly assumed that private trade did not reach particular communities; and free seed distribution was provided when many local farmers were accessing these inputs already from private channels (Longley et al, 2001). However, since 1991, important lessons have been learned about what makes a programme appropriate and effective. Some of these are outlined below.

Cash-based relief

Among the most important lessons learned is that injections of small amounts of cash to poor households and individuals can make an important difference. With well-developed food markets in most of the Jubba region, ICRC has used emergency cash injections to help needy families purchase food and other critical commodities, and these also have had generally favourable impacts. Cash was given 'in return for their labour in rehabilitating vital community infrastructure, usually a rainwater catchment or irrigation channel' (ICRC, 2004). Outside the Jubba area the Emergency Cash Relief Programme funded by Oxfam-UK and Novib (a Netherlands NGO) provided cash to drought-affected households in the Sool Plateau area of northern Somalia, a programme that had several advantages over commodity-based relief.[8]

Relationship between food security responses and early warning systems

Not surprisingly, many externally-funded interventions in the Jubba area have focused on early warning systems for assessing emergency food and other humanitarian needs. These have been based outside of Somalia (in Nairobi) and rely on a range of data sets, including remotely sensed information. Early warning systems used by FSAU and USAID-funded FEWS NET draw on a range of indicators that rank locations in terms of stages: famine, humanitarian emergency, acute food and livelihood crisis, alert, and non-alert or 'normal'. The levels are differentiated by recorded rates of malnutrition and child mortality, food availability, asset depletion, civil conflict and other important variables. In southern Somalia, the livelihood classification system is used to calculate the number of people in a particular livelihood zone that require assistance based on immediate humanitarian needs and/or because of a livelihood crisis; the latter is normally signalled by unusually high patterns of out-migration and asset depletion (Hemrich, 2005).

Based on the livelihood classification system discussed earlier, rainfall data and seasonal crop estimates, the FSAU system calculates food deficits per livelihood/food economy zone. The estimated deficit is the gap between what is required to meet the population's basic nutritional requirements and what is expected to be produced and purchased. These estimates are figured into the consolidated annual appeal to international agencies for assistance.

FSAU has gone further than other groups in developing a systematic classification system for assessing food security and humanitarian needs in southern Somalia and elsewhere. The system is called the 'Integrated Food Security and Humanitarian Phase Classification' (FAO/FSAU, 2006). It is a general tool for classifying regions according to five broad categories, ranging from '(1) generally food secure' to '(5) famine/humanitarian catastrophe'. The system is not a substitute for fine-tuned sectoral or field-based analyses, which have much greater capacity for refined analyses of site- or sector-specific interventions. But it has the advantage of being based on a livelihoods approach that allows interested parties to assess the condition of different livelihood systems and the potential impacts that a planned programme or policy will have on them. As of early 2007 it was too soon to tell whether development and humanitarian agencies would adopt the system and how effective it would be for identifying humanitarian needs and strengthening the links between food security and livelihoods analysis and action.

Longer-term development planning and policy

What would be entailed in a policy framework for a country and region (Jubba area) that currently has no representative formal institutions to make or implement policies? Unlike in a sovereign nation, a policy framework in this case would be of primary utility to NGOs and donors and the communities with whom they work. It would need to be implemented through these organizations, especially those that represent specific communities, until representative national and regional institutions are formed in Somalia. The danger in any situation where there are weak or non-existent local and regional institutions is that the policy agenda can be driven to too great an extent by external agencies; unfortunately the perception of many Somalis about current assistance programmes in the country is that they have been externally driven. Remarks such as 'How can NGOs know about the needs of southern Somalia when most of their staff rarely visit the country?' are sometimes heard.

At a minimum a so-called enabling environment for longer-term development requires considerably better physical security than exists now in Somalia – a situation that does not put the community and/or development personnel at such high risk. In fact, development investments can actually aggravate political conflicts if caution is not used. Questions that need to be asked of any type of programme in the area include: (1) how will political factors affect the programme?; and (2) how will the programme itself impact existing political and potentially conflictive relations in the area? As earlier discussions showed, there was almost no three-month period from 2000 to 2005 during which some part of the region was not experiencing armed conflict (see Table 6.1).

A policy framework for protracted emergencies should attempt to build on many of the same positive elements that would be included in any rural development policy. The EC is attempting to do this in its five-year strategy and in some of its programmes (EC, 2002; 2006). These entail strong participation of the local community, a sense of local ownership and strong collaboration with local institutions. Again, these will not be straightforward in the Jubba area because questions of who to collaborate with can be highly political and potentially conflictive. Once again, there needs to be sufficient social and political analyses to know which institutions represent the local community, and what kinds of political risks are associated with supporting certain groups vis-à-vis others. Unfortunately, humanitarian agencies without adequate knowledge have worked with local groups that represented militia factions rather than households and communities.

Since 1991 we have learned much about what does not help sustained development in the Jubba area. These include interventions that: (1) reduce the flexibility of household coping and recovery strategies; (2) compete or bypass already existing market channels; (3) require local institutions to do too much during implementation; and (4) heighten political conflict and insecurity. Much of the livelihoods-related relief work in the Jubba area has been related to animal health and has attempted to develop programmes based on local participation and principles of sustainability. There has been attention to developing a system of animal health delivery that is based on user fees and training of CAHWs that could sustain itself without external assistance (Catley, 1999; PACE, 2004). Overall the livestock sector is perhaps the best example in the Jubba region of utilizing data to address food security problems in the medium to long term with the goal of moving from a relief/humanitarian to a development mode.

Importantly, some of this work has been planned and implemented with active participation from local communities, traders and veterinarians (see Bishop et al, this volume). Such experience has shown that seeking local participation in problem identification, planning and implementation, involving key stakeholders and building on local knowledge systems and practices are as important for successful projects in protracted emergency situations as they are in stable environments.

Nonetheless, any policy deliberations must start with the reality that the area will continue to require considerable amounts of short-term humanitarian assistance and much of this will continue to be based in the riverine areas and camps for IDPs. There are several areas, however, where social data are critical for understanding both immediate humanitarian needs and longer-term food security responses. One critical area that has lagged behind others is the collection of gender-disaggregated data, which has been constrained by the volatile security situation. For longer-term programmes of food security and nutrition, gender-disaggregated data is absolutely essential, especially for understanding how interventions might affect the welfare status of vulnerable household members. We know that a large proportion of those in IDP camps in Somalia and refugee camps in Kenya are women and children (Little, 2003), and that some of the most nutritionally at-risk populations tend to be women and children. This would imply that they have borne the brunt of the hardships in Somalia and a longer-term development approach must address their situation.

It goes without saying that a policy framework for the Jubba area must be flexible, monitored and context-specific. Longer-term development programmes face many unknowns in a protracted political crisis and policy guidelines for directing programme investments in the Jubba area need to be extremely flexible. Nonetheless, the development environment has not been as stagnant since 1991 as some observers might believe. Certain Somali communities and households have taken advantage of new market opportunities (such as cross-border trade with Kenya), rebuilt assets, diversified production systems, developed informal institutions (including money transfer companies) and so on. In short, certain developments have proceeded in the region regardless of whether outsiders have recognized them.

Effective flexibility means timing relief and development assistance to periods and locations when it is most needed. In a difficult environment like Jubba, impacts are strongly determined by timing issues. At the same time development programmes need to be cognizant of the politics of any type of development interventions and insure that their timing does not aggravate or rekindle political conflicts.

Conclusions

This chapter has shown how household assets and market access are key factors in reducing poverty and vulnerability in the Jubba region. It supports the general findings of the work of Longley et al in southern Somalia (2005): 'Longer-term aid strategies can only be effective when households are in a position to rebuild and stabilize their asset bases. ... An important aspect of the dimension of stability relates to the stability of food access and... this requires the accumulation of *ownership and access to assets* by households' (emphasis added). The key for longer-term development planning is to protect existing economic assets, rehabilitate damaged physical assets (for example, irrigation infrastructure), and develop through skills training and education those human capital assets necessary for the future. This can be done through carefully planned programmes – in areas like preventive health, primary education or community-based irrigation rehabilitation – that work closely with local communities and build on their needs.

The chapter also considered what might be entailed in a longer-term development approach for the area and how this relates to current information and institutional needs. For a longer-term policy framework it is important to know more about who the food insecure are, how the poor versus non-poor categories change proportionately vis-à-vis shock/recovery cycles, and what are the asset poverty and food insecurity thresholds for different livelihood zones in the Jubba area. The chapter suggested an approach for looking at the dynamics of poverty and food insecurity in the area. What appears to be lacking in current information systems is analyses of the required asset bundles that households and individuals require to escape from poverty and food deprivation, recover from shocks (ecological, economic and political) and/or avoid falling into poverty and food insecurity. As noted, in a highly volatile environment like the Jubba region, households over time may move in/out of poverty. The type of analysis suggested would support development agencies in devising strategies that help to rebuild household and community assets to levels that are sustainable in the medium to long term (5–10 years). Nonetheless, any longer-term development policy framework must recognize that short-term humanitarian needs will continue to require considerable attention and resources for the foreseeable future.

As this chapter has shown, there are considerable economic and social developments already taking place in parts of the Jubba in the absence of programmes of planned development. With a vacuum in systematic data collection that has lasted more than 15 years, any investments in longer-term development will encounter unknowns and may have unintended consequences. The key is to minimize the negative effects of these. A longer-term development agenda for complex emergency areas like Jubba requires considerable innovation in programmes and institutional relationships. Not unexpectedly, these will entail some level of risk and continued gathering of information is needed to insure that they benefit Jubba communities and residents. With some minimum prospects for a political solution in 2007, it is important that attempts be made to consider longer-term policy and development needs in the area, instead of just humanitarian assistance.

Acknowledgements

There are several individuals and organizations that assisted me in assembling the background materials for this chapter. They include Nick Haan and Carol King'ori of the Food Security Analysis Unit (FSAU)-Somalia, M. Aw-Dahir of the Famine Early Warning Systems Network (FEWS-Net), Philip Steffen of USAID, Hussein Mahmoud of Egerton University (Kenya) and the Max Planck Institute for Social Anthropology (Germany), Riccardo Costagli of Terra Nuova, Friedrich Mahler of the European Commission (EC)-Somalia Unit, and Andy Catley and Tim Leyland of Tufts University. I also am grateful to Luca Russo, Luca Alinovi and Guenter Hemrich of the Food and Agriculture Organization of the United Nations (FAO) who provided initial guidance on how to approach this study and the writing of the chapter. As the author, however, I take sole responsibility for its contents and none of the above individuals, institutions and agencies should be held accountable.

References

Ali, D., Toure, F. and Kiewi, T. (eds) (2005) 'Cash Relief in a Contested Area: Lessons from Somalia', Humanitarian Practice Network Paper No. 50, Overseas Development Institute, London.

Barrett, C. B. and Maxwell, D. (2005) *Food Aid After Fifty Years: Recasting Its Role*, Routledge, London.

Campbell, E. H. (2005) *Formalizing the informal economy: Somali refugee and migrant trade networks in Nairobi*, Global Migration Perspectives No. 47, Global Commission on International Migration, Geneva.

Carter, M. and Barrett, C. (2006) 'The economics of poverty traps and persistent poverty: An asset-based approach', *Journal of Development Studies* 42 (2): 178–199.

Carter, M., Little, P. D., Mogues, T. and Negatu, W. (2004) *Tracking the Long-Run Economic Impacts of Disasters: Environmental Shocks and Recovery in Ethiopia and Honduras*, BASIS CRSP, Department of Applied and Agricultural Economics, University of Wisconsin, Madison, WI.

Catley, A. (1999) *Community-Based Animal Health Care in Somali Areas of Africa: A Review*, PanAfrican Rinderpest Campaign (PARC), Organization of African Unity (OAU)–Interafrican Bureau for Animal Resources (IBAR), Nairobi.

Chronic Poverty Research Centre (2004) *The Chronic Poverty Report, 2004–05*, Institute for Development Policy and Management, University of Manchester, Manchester.

De Waal, A. (1997) *Famine Crimes*, James Currey Publishers, Oxford.

EC (European Community) (2002) 'Strategy for the implementation of special aid to Somalia (SISAS)', EC, Brussels, www.delken.ec.europa.eu/en/information.asp [Accessed 16 August 2006].

EC (2006) *Livelihoods Support Programme in South Somalia*, EC, Brussels.

Evans, H., Cullen, M. and Little, P.D. (1988) *Rural–Urban Exchange in the Kismayo Region of Somalia*, Cooperative Agreement on Settlement and Resource Systems Analysis, Worcester, MA and Binghamton, NY.

FAO/FSAU (2006) *Integrated Food Security and Humanitarian Phase Classification. Technical Manual, Version I*, Technical Series Report No. IV(11), FAO/FSAU, Nairobi.

FSAU (Food Security Analysis Unit-Somalia) (undated) *Livelihood Baseline Profile: Lower and Middle Jubba Agropastoral*, FSAU, Nairobi.

FSAU (2004) Map of Somalia's livelihood zones, FSAU, Nairobi.

FSAU/UNICEF (2004) *Jilib Riverine Nutrition Survey, Middle Jubba Region, Somalia*, FSAU, Nairobi.

Hemrich, G. (2005) 'Matching food security analysis to context: the experience of the Somalia Food Security Assessment Unit', *Disasters* 29 (s1): s67–s91.

Hoddinott, J. and Quisumbing, A. R. (2003) *Methods for Microeconometric Risk and Vulnerability Assessments: A Review with Empirical Examples*, Social Protection Discussion Paper No. 0324, World Bank, Washington DC.

ICRC (International Committee of the Red Cross) (2004) *Annual Report*, ICRC, Geneva.

Little, P. D. (1989) *The Livestock Sector of the Kismayo Region, Somalia: An Overview*, Working Paper No. 50, Institute for Development Anthropology, Binghamton, NY.

Little, P. D. (1992) 'Traders, brokers, and market "crisis" in southern Somalia', *Africa* 62 (1): 94–124.

Little, P. D. (2003) *Somalia: Economy Without State*, James Currey Publishers, Oxford, UK and Indiana University Press, Bloomington, IN.

Little, P. D. (2005) *'Unofficial trade when states are weak: the case of cross-border commerce in the Horn of Africa'*, Research Paper No. 2005/13, World Institute for Development Economics Research, United Nations University, Helsinki.

Little, P. D., Smith, K., Cellarius, B. A., Coppock, D. L. and Barrett, C. B. (2001) 'Avoiding disaster: Diversification and risk management among East African herders', *Development and Change* 32 (3): 401–433.

Little, P. D., with assistance from Abdel Ghaffar, M. A., Carter, M., Roth, M. and Negatu, W. (2002) 'Building assets for sustainable recovery and food security', BASIS Brief No. 5, BASIS Research Program, Department of Agricultural and Applied Economics, University of Wisconsin, Madison, WI.

Little, P. D., Stone, M. P., Mogues, T., Castro, A. P. and Negatu, W. (2006a) 'Moving in place: drought and poverty dynamics in South Wollo, Ethiopia', *Journal of Development Studies* 42 (2): 200–225.

Little, P. D., McPeak, J., Barrett, C. and Kristjanson, P. (2006b) 'The multiple dimensions of poverty in pastoral areas of East Africa: An overview paper', presented at the 'Conference on pastoralism and poverty reduction in East Africa: A policy research conference', International Livestock Research Institute (ILRI), Nairobi, 27–28 June.

Longley, C., Hemrich, G. and Haan, N. (2005) 'Reconfiguring aid: Food security programming and policy in Somalia', FAO, Rome (mimeo).

Lybbert, T. J., Barrett, C. B., Desta, S. and Coppock, D. L. (2004) 'Stochastic wealth dynamics and risk management among a poor population', *Economic Journal* 114: 750–777.

McPeak, J. and Little, P. D. (2004) 'Cursed if you do, cursed if you don't: The contradictory processes of sedentarization in northern Kenya'. In E. Fratkin and E. Roth (eds) *As Nomads Settle: Social, Health and Ecological Consequences of Pastoral Sedentarization in Northern Kenya*, pp. 87–104, Kluwer Academic/Plenum Publishers, New York.

Menkhaus, K. (1999) 'Studies on governance in Lower Jubba Region', UNDOS, Nairobi.

Menkhaus, K. (2003) 'Somalia: a situation analysis and trend assessment', UNHCR, Geneva.

Menkhaus, K. and Craven, K. (1996) 'Land alienation and the imposition of state farms in the lower Jubba Valley'. In C. Besteman and L. Cassanelli (eds) *The Struggle for Land in Southern Somalia: The War behind the War*, pp. 133–153, Westview Press, Boulder, CO.

Narbeth, S. and McClean, C. (2004) 'Livelihoods and protection: Displacement and vulnerable communities in Kismaayo, southern Somalia', Humanitarian Practice Network Paper 44, Overseas Development Institute, London.

PACE (PanAfrican Programme for the Control of Epizootics) (2004) *Final Report: PACE-Somali Component, 1st Phase, October 2001–April 2004*, Inter-African Bureau for Animal Resources/Africa Union (IBAR/AU), Nairobi.

Prendergast, J. (1997) *Crisis Response: Humanitarian Band-Aids in Sudan and Somalia*, Pluto Press, London.

UNDP (United Nations Development Programme) (2001) *Human Development Report: Somalia*, UNDP, Nairobi.

UNDP/World Bank (2003) *Socio-Economic Survey: Somalia, 2002*, Report No. 1, Somalia Watching Brief, UNDP/World Bank, Nairobi.

UNOSOM (United Nations Operation in Somalia) (1994) 'Report on the Lower Jubba Peace and Reconciliation Conference', unpublished paper, UNDOS Documentation Unit, Nairobi.

World Bank (2005) *Somalia: From Resilience Towards Recovery and Development*, Report No. 34356-SO, World Bank, Washington DC.

CHAPTER 7

Livestock and livelihoods in protracted crisis: The case of southern Somalia

Suzan Bishop, Andy Catley and Habiba Sheik Hassan

Abstract

This chapter outlines the role of livestock in human food security and livelihoods in Somalia, and analyses the livestock interventions of both relief and development programmes since the end of the civil war in 1991. Focusing on the more conflictive central and southern parts of the country, it examines the extent to which relief and development approaches were harmonized and based on livelihoods thinking and information.

Introduction

This chapter provides an overview of livestock and pastoral livelihoods in Somalia, and analyses the livestock interventions of the major relief and development programmes. It explores the approaches used, the technical rationale for the interventions and the impacts achieved, and discusses the conditions and mechanisms for coordinating livestock programmes. Central to this analysis is the extent to which aid agencies engaged Somali communities and stakeholders in the joint analysis of problems related to livestock and to programme design, implementation and evaluation. The chapter draws on a literature review, baseline livelihoods profiles and other recent information available from the FSAU for Somalia (FSAU Somalia, 2006a) and original research by a team that included the authors.[1]

Pastoral livelihoods in Somalia: Livestock and beyond

The economy of Somalia is often described as dependent on livestock and this is in many ways true. Around 50 per cent of Somalia's 637,600 km^2 is arid or semi-arid land, and pastoral livestock production has long played a crucial role in people's livelihoods. Somalia has relatively more people involved in pastoralism and more land used for pastoralism than any other country in the Horn of Africa region. Like other pastoralists, Somali pastoralists rely on livestock not only for food, income and transport, but also as the basis for many social transactions and to some extent, for social organization. However, there are important aspects of Somalia's economy and of its pastoral groups that distinguish it from neighbouring countries in the Horn of Africa region.

Somalia possesses the longest coastline of any country in Africa, and since the 1830s Somalis have travelled overseas to find work and send money home to relatives (Geshekter, 1993). Due to links with Arab traders and merchants, Somalis regularly travelled to the Gulf States and during the colonial period were employed as sailors and other kinds of workers. Pilgrimages to Islamic centres also helped to ensure that Somalis were not isolated from news and experiences from other countries. Economic prosperity in the oil-producing countries of the Gulf has attracted Somalis since the 1970s, and by 1986 there were an estimated 300,000 Somalis in the United Arab Emirates and Saudi Arabia (Janzen, 1986).

This mobility has provided many Somalis with economic alternatives beyond what is available within Somalia. In 1987 the International Labour Organization (ILO) estimated remittances sent to Somalia at $280 million (Geshekter, 1993). The ILO characterized Somali families as multi-occupational, multi-national production units; a family grazing their livestock on the Ethiopian border could receive significant support from relatives abroad via the clan system that extended even overseas. By 2006 the Somali diaspora was estimated at over 1 million people and remittances had reached figures estimated at between $825 million per year, or around 60 per cent of GDP (Economist Intelligence Unit, 2006) and $1 billion (Lindley, 2005). FSAU livelihoods surveys indicated that remittances accounted for up to 20 per cent of the income of poor households in some pastoral livelihoods zones (see Table 7.1).

Another unusual feature of the Somali economy is that pastoralism has supplied the bulk of the livestock for the country's dynamic and well-established livestock export trade, and for decades Somali pastoralists have oriented their production toward commercial purposes. During the protracted crises in the country, livestock markets continued to function and adapted to changing pressures and opportunities. Somali pastoralists are also involved in petty trade, charcoal production, the harvesting and sale of frankincense and other activities.

Throughout the Horn of Africa, there are examples of pastoral livelihoods that rely heavily on seasonal cross-border movements. For many of these groups, their home territory falls within, and is partly defined, by national boundaries. However, for Somali pastoralists the traditional clan territories transgress the Somali border for virtually its entire length, from northeast Kenya to northeast Djibouti. These communities live on both sides of the border, rather than moving back and forth across it. Pastoral livelihoods, especially in southern Somalia, are closely linked to events, services and programmes in Kenya and Ethiopia.

Table 7.1 Examples of livelihood baseline profiles for southern Somalia

Livelihoods zone	*Sources of food*		*Sources of income*	
Southern Inland Pastoral (parts of Bakol, Bay, Gedo, Hiran, Lower and Middle Juba, Lower and Middle Shabelle regions)	***Poor households***		***Poor households***	
	Livestock products	39%	Milk and dairy sales	80%
	Staple purchase	52%	Livestock sales	10%
	Non-staple purchase	5%	Gums and resins	10%
	Gifts	4%		
	Middle-income households		***Middle-income households***	
	Livestock products	66%	Milk and dairy sales	75%
	Staple purchase	26%	Livestock sales	23%
	Non-staple purchase	8%	Gums and resins	2%
Addun Pastoral (central areas of Mudug, Nugal and Galgadud regions)	***Poor households***		***Poor households***	
	Livestock products	15–20%	Livestock sales	35–45%
	Purchase	70–75%	Self-employment	20–25%
	Gifts	10–20%	Employment	20–25%
			Gifts and remittances	10–20%
	Middle-income households		***Middle-income households***	
	Livestock products	25–35%	Livestock sales	85–90%
	Purchase	65–75%	Labour	0–10%
			Remittances	0–10%
Hawd Pastoral (western edge of central Somalia and southern edge of Somaliland)	***Poor households***		***Poor households***	
	Livestock products	25–35%	Livestock, livestock products	40–45%
	Purchase	50–60%	Employment	35–40%
	Gifts	10–15%	Gifts and remittances	10–15%
	Wild foods	0–5%	Wild foods	5–10%
	Middle-income households		***Middle-income households***	
	Livestock products	45–55%	Livestock, livestock products	70–80%
	Purchase	45–55%	Self-employment	10–15%
	Wild foods	0–10%	Water sales	5–10%
			Remittances	0–10%
Central Agropastoral (eastern areas of Galgadud, Middle Shabelle and Mudug regions)	***Poor households***		***Poor households***	
	Livestock products	10–20%	Livestock, livestock products	45–55%
	Crops	25–35%	Self-employment	25–30%
	Purchase	35–45%	Community support	5–15%
	Wild food and gifts	10–20%	Other sources	5–15%
	Middle-income households		***Middle-income households***	
	Livestock products	35–45%	Livestock, livestock products	70–80%
	Crops	20–30%	Self-employment	0–10%
	Purchase	30–40%	Other	15–25%

Source: FSAU Somalia (2006a)

Livestock assets, food and income

Macro-level indicators

Somali pastoralists often keep mixed herds of livestock as a strategy for reducing the risk of losses due to drought or disease, and to meet market demands for specific types of livestock. Between 1986 and 1990, the national herd was estimated at 42.1 million animals, comprising 31.2 million sheep and goats, 6.3 million camels and 4.6 million cattle (Baumann, 1993; more recent reports repeated or adapted the data from the 1980s). Pastoral livestock production in Somalia is thought to contribute approximately 60 per cent of the income or subsistence of the population and 80 per cent of total exports (FAO/World Bank/EU, 2004).

The modern history of the Somali livestock export trade starts with the oil boom in the Gulf States in the early 1970s and a growing demand for live animals, particularly during the Hadj religious holiday period. Somali livestock were highly appreciated by Arab consumers, and the trade soon expanded to reach a record pre-war level in 1982, when 1.45 million sheep and goats, 157,000 cattle and 15,000 camels were exported (Reusse, 1982; Janzen, 1986). The robust and flexible nature of the Somali livestock export business was demonstrated by its response to a Saudi Arabia ban on Somali cattle exports in 1983 due to rinderpest. Traders in southern Somalia moved cattle across the border to the Garissa market in northeast Kenya, where by 1998 more than 100,000 Somali cattle were being sold (Little, 2003 and this volume).

From 1988 to 1991, the livestock export trade was interrupted by the civil war. However, substantial unregulated trade then developed that soon exceeded pre-war levels. The trade peaked in 1997 when 3.13 million sheep and goats, 70,733 cattle and 57,946 camels were exported from Berbera and Bossasso (FAO/World Bank/EU, 2004). Although Saudi Arabia imposed bans on livestock imports from the Horn of Africa in 1997 and 2000 (and the latter ban remained in force in 2006), by 2002 live animal exports had recovered to 57 per cent of the pre-ban numbers for sheep and goats, and 52 per cent for camels. In 2002 there was actually a 27 per cent increase in cattle exports relative to pre-ban figures. Somalis also responded to the Saudi bans by exporting chilled meat (which was permitted into Saudi Arabia) rather than live animals. By 2006, eight modern slaughterhouses were filling an increasingly important niche market based on the air freight of freshly slaughtered carcasses of young sheep and goats to the Gulf.

When examining Somali livestock export figures it should be noted that many animals shipped from the northern ports of Berbera and Bossasso originate in the Somali region of Ethiopia. The informal cross-border livestock trade from Ethiopia to Somalia is difficult to quantify, but some studies attribute up to two-thirds of Somali livestock exports to Ethiopian origin (Stockton and Chema, 1995). For southern Somalia, the key export trade has been the cross-border trade of cattle to Kenya.

Household-level indicators

The FSAU livelihoods baseline surveys clearly show the importance of livestock for households in southern Somalia. Four main livelihoods zones (in terms of geographical coverage) are summarized in Table 7.1 as examples.

Within specific ecosystems and livelihoods zones in southern Somalia, there are notable differences in herd composition. The central rangelands typify Somali pastoralism, with relatively high camel, sheep and goat ownership and few cattle (for example, the Addun Pastoral Zone) (see Table 7.2). In contrast, cattle ownership increases further south (for example, the Southern Inland Pastoral Zone) (see Table 7.2) and is particularly high in the Lower Jubba region.

The mixed herd composition shown in Table 7.2 not only reflects the physical environment of the zone and pastoralists' risk aversion strategies, but also market demand. In the central rangelands, sheep and goats are the most important livestock species for generating cash and herds are managed so that they supply young male animals to domestic and export markets (Nauheimer, 1993). Prior to the civil war, Somali law allowed the export of male animals only, as a strategy for maintaining the national breeding herd. Pastoralists changed the composition of herds in response to changing markets. Working in the Bay region of Somalia, Al-Najim (1991) attributed dramatic increases in cattle ownership in part to a demand for live cattle in Saudi Arabia. The commercial aspects of pastoralism in Somalia are not restricted to animal sales but also include livestock products, particularly milk, and rural to urban milk marketing arrangements are well organized around urban centres such as Mogadishu (Herren, 1990), Kismayo and Afmadow (Little, 2003) and Mandera in northern Kenya.

Given the importance of livestock to pastoral households, herd ownership reflects wealth. From the perspective of livelihoods analysis, household livestock assets can be viewed as a key measure of people's vulnerability and capacity to withstand shocks. Compared to Borana pastoralists in southern Ethiopia and Maasai in Tanzania and Kenya, Somali cattle herders in Lower Juba have higher livestock holdings and are considered to be relatively wealthy (Little, 2003). However, FSAU reports indicate that in some areas Somali

Table 7.2 Examples of livestock holdings by wealth group in southern Somalia

Wealth groups, by livelihoods zone	*Livestock ownership*			
	Camels	*Cattle*	*Sheep, goats*	*Donkeys*
Southern Inland Pastoral				
Very poor (0–5%)	5–15	2–4	15–25	–
Poor (15–30%)	20–30	5–10	30–50	0–1
Middle (40–50%)	40–60	15–25	60–90	1–2
Better-off (25–35%)	70–100	30–40	100–250	2+
Addun Pastoral				
Poor (25–30%)	2–5	–	40–60	–
Middle (45–55%)	10–15	–	80–120	–
Better-off (20–25%)	25–30	–	150–200	–
Hawd Pastoral				
Poor (20–30%)	5–10	0	50–60	0–1
Middle (45–55%)	25–30	10–15	80–100	–
Better-off (15–20%)	25–30	30–40	200–250	–
Central Agropastoral				
Poor (20–30%)	1–5	0–3	20–40	–
Middle (50–60%)	5–15	3–5	50–70	–
Better-off (15–25%)	20–40	5–10	80–150	–

Source: FSAU Somalia (2006a)

pastoralists are becoming poorer, with more households shifting from middle to poor wealth groups, and from poor to very poor wealth groups; these trends are described in more detail in the next section.

The impact of trends and shocks on pastoralist livelihoods

Pastoralists in southern Somalia have experienced a complex mix of trends and shocks since the end of the civil war in 1991. Some important trends were set in motion in the pre-war period and included the emergence of the livestock export trade to the Gulf. As trade developed, a class of wealthy herders and urban-based livestock traders appeared who started to appropriate key grazing areas and encouraged the production of small ruminants over other livestock types. Hence, traditional pastoralism began to change as less powerful herders were diverted from their traditional grazing areas and faced increasing competition for water and other resources. In some areas, new settlements began to appear that were associated with agropastoralism, more accessible water points and increased ownership of cattle. These trends are well documented in studies such as those conducted by Al Najim (1991) in the Bay region.

Since the early 1990s, various forms of conflict and the breakdown of government and regulation have added further complexity to the pre-war trends of sedentarization, rangeland enclosure and expansion of water points. There are reports of expanded private enclosure of land, construction of more private water reservoirs (*berkhads*), unregulated exploitation and damage to natural resources such as cutting of trees and charcoal production, and further marginalization of more vulnerable households (Bradbury et al, 2001). These trends, along with human population growth, may be associated with increasing numbers of destitute former pastoralists in urban areas.

At the same time there was a gradual emergence of local public administrations, largely independent of central government and comprising various forms of local representation such as traditional elders' councils, women's associations and Sharia courts. As these relatively representative and participatory modes of local government evolved in different areas, there were indications that local regulations and socio-cultural norms might lead to strategies for better provision of services and more sustainable development. Unfortunately for southern Somalia, these institutions were relatively weak compared to those in Somaliland and Puntland, and in 2001 it appeared that political elites remained in power (Bradbury et al, 2001).

Somali pastoralists have experienced four main shocks since the early 1990s: conflict, drought, floods and livestock disease outbreaks, which in some cases were associated with livestock export bans (see Box 7.1). Livestock bans resulted in both short-term shocks and longer-term impacts on livelihoods such as reduced livestock prices. These shocks have impacted on pastoralist livelihoods and in particular on livestock holding. Using the ratio of TLU per African adult male equivalent (AAME), Little compared livestock holdings in

Box 7.1 Livestock-related shocks to pastoral livelihoods in southern Somalia

Conflict

Conflict and episodes of food insecurity are strongly linked. Conflict causes loss of human life, injury, displacement, loss of assets and abandonment of productive activities. Conflict also inhibits free movement of animals and goods, and reduces access to markets, water and pasture. Service provision and trade activities are harmed due to increased risks and higher transaction costs. Social capital is eroded and inter-clan relationships are weakened, leading to less efficient joint use of natural resources.

Drought

Droughts are frequent in Somalia and have devastating consequences. Drought causes major social and environmental shocks to pastoral systems, with loss of animals, long-term shifts in herd structure from cattle to small ruminants and increased settlement with potential environmental degradation. Drought effects are cumulative, with poor-quality stock carried, leading to slowed herd growth. Forced liquidation at depressed prices decimates herds. Animal markets become unstable and can collapse. The combination of high grain prices and depressed livestock prices weakens the purchasing power of pastoralists and exposes them to severe food insecurity.

Floods

Floods are sporadic and less common than droughts. Destructive floods occurred during the 1996–1999 El Niño, destroying herds, homes and livelihoods, and causing outbreaks of livestock disease. Flooding can also damage roads and bridges, disrupting trade. In late 2006 major flooding and humanitarian crisis in the Shabelle and Juba river basins affected up to 1 million people.

Livestock diseases

Livestock disease outbreaks can affect pastoral livelihoods in Somalia in two main ways:

Livestock mortality and direct production losses. Disease can cause death of livestock and important loss of production. Diseases with high mortality and livelihoods impact include contagious caprine pleuropneumonia and peste des petits ruminants (goat plague).

Bans on livestock exports. Livestock export bans imposed by Somalia's trading partners result in rapid decreases in livestock prices. Bans by Saudi Arabia related to Rift Valley fever led to a short-term 40–50 per cent in livestock prices and 90 per cent fall in livestock sales, and some prolonged depression of prices. Pastoral incomes were markedly reduced and the loss of revenue from exports led to fewer food imports being purchased; prices rose by 60–100 per cent. The technical basis for Saudi-imposed bans has often been questioned because Saudi sanitary requirements have been both vague and inconsistent.

the Afmadow district of Lower Jubba in 1988 and 2005. Whereas in 1988 the average TLU to AAME ratio was 8.5 and the middle two quartiles averaged 6.0, in 2005 84 per cent of the population had livestock holdings of less than 2.7 TLU to AAME.[2] More recent evidence for decreasing livestock holdings in southern Somalia is available from FSAU assessments conducted in September 2006 (FSAU Somalia, 2006b), which attribute decreasing livestock numbers to the drought in 2005–2006, conflict and reduced access to pasture and water (see Table 7.3). In terms of vulnerability, these trends were most likely to impact on poor or very poor households who had less livestock than wealthier households.

Table 7.3 Trends in livestock holdings in southern Somalia, April 2005 to March 2006

Region	*Livestock holdings*		
	Camels	*Cattle*	*Sheep and goats*
Gedo	5–10% decrease	40–60% decrease	30–50% decrease
Juba Valley	0–5% decrease	40–55% decrease	15–30% decrease
Bay	5–14% increase	15–25% decrease	0–15% decrease
Bako	0–10% increase	15–20% decrease	0–15% decrease
Shabelle Valley	No change	No change	No change
Hiran	No change	No change	No change
Galgadud and south Mudug	0–15% increase	0–5% decrease	0–5% decrease

Source: FSAU Somalia (2006b)

Other evidence of this downward trend includes increased numbers of pastoralist dropouts, proportional shifts in numbers toward lower wealth groups, increased trekking distances to water and pasture, environmental degradation and reduced mobility. The variable and often harsh conditions in south and central Somalia mean that mobility is critical to enhancing resilience, managing risk and being able to manage critical safety net areas of dry season rich patches of vegetation, many of which are being increasingly lost to agriculture (irrigated and rainfed) or conflict. The Gedo region is particularly affected by reduced mobility, with increased drought-related mortality in herds reported more regularly in this region relative to neighbouring areas.

Adapting to the crisis

Despite the shocks described above, pastoralist households in Somalia have demonstrated significant capacity to adapt to the crisis.

Livestock and social capital

Livestock has huge socio-cultural importance in traditional pastoral societies, and Somali pastoralists are particularly dedicated to their animals. The rationale for accumulating larger herds in good times should be viewed not only in terms of risk mitigation and insurance against hard times, but also as a form of social capital. The social relevance of livestock wealth is perhaps best illustrated by reference to the social organization of Somalis and in particular, the *dia*-paying group. Despite the widespread cultural and religious homogeneity of Somali society, the population is deeply divided along clan and subclan lines. Clans are highly influential in all aspects of political, economic and social life. They are sources of conflict and polarization, but are also the main sources of conflict resolution, mediation and negotiated settlements. The smallest but probably most robust clan unit is the *dia*-paying group, defined as a group of related males who bear collective responsibility to pay compensation for the killing of a member of another *dia*-paying group. Traditionally, this compensation is defined in terms of livestock, so that livestock holdings have a direct bearing on clan influence and power.[3]

In the face of the harsh Somali environment and the repeated depletion and rebuilding of livestock assets due to drought or disease, a high degree of social collaboration and reciprocity helps Somali communities to manage livestock efficiently in terms of labour requirements for herding and watering animals, and herd reconstitution. Strong social support systems also enable collective assistance to households in particular need, for example, those experiencing severe loss of livestock (see Box 7.2).

Local organization and knowledge

Somali communities also have a long history of organizing themselves to take collective action and provide themselves with sustainable services. One of the best examples of a sustainable traditional service in Somali areas is the system of Koranic education, based on mobile schools or *duksis* that follow the herds. Early descriptions of this service date back to the 1960s (Lewis, 1961) and it was still much in evidence in pastoral areas 35 years later (Pearson, 1996). Before the civil war the Ministry of Education realized that conventional education services based on modern, fixed schools were largely unworkable for pastoralist communities and sought to integrate Koranic schooling into the official education programme (Gorham, 1978). Somali communities have

Box 7.2 Traditional Somali systems for provision of livestock to poor families

Free gift (xologoyo)

Livestock given freely to a needy family. A committee of elders is organized to collect livestock from relatives of the recipient family; the number of animals provided depends on whether the family is expected to engage in farming or herding activities.

Loan (maalsin)

Usually arranged between two individuals and involving lactating cattle or goats. The borrower returns the original stock to the lender with any offspring when the animals give birth, or keeps the offspring and returns only the adult females to the lender. The terms of the loan depend on kinship ties between the two parties.

Marriage

When a poor man with no livestock marries, his relatives give him livestock (no set types or quantities). When a woman marries a poor man, her relatives – mostly father, brothers and uncles – provide her with livestock when she visits her father's family.

Almsgiving

Almsgiving related to Somali customs and livestock provision requires those people who hold a certain number of livestock to provide animals to poor families as follows:

Livestock type owned	***Number in herd***	***Livestock to be donated***
Camel	5 or more	1 sheep, 2 years old
Cattle	30 or more	1 calf, 2 years old
Sheep/goats	40 or more	1 sheep, 2 years old

A man with camels is not obliged to give alms until he owns five. As the herd of an almsgiver increases, so does the number of animals donated (a man with 15 camels would provide three sheep). Almsgiving occurs once a year and is coordinated by a committee of elders who collect livestock from the alms providers.

Source: Catley, 1999

also organized themselves for natural resource management (Bradbury et al, 2001) and provision of primary health services (Bentley, 1989).

Somali pastoralists are known among pastoralist communities in the Horn of Africa for their livestock-rearing skills and knowledge of animal diseases. There have been reports of this knowledge since the 1950s, including accounts of local disease terminology and traditional quarantine and vaccination practices (Mares, 1951; 1954). Research in the early 1990s indicated that extensive local livestock knowledge and skills were still very much in evidence (Catley and Walker, 1997).

Movement and seasons

Arid and semi-arid areas are characterized by high variability of rainfall in terms of both the volume of rain and its spatial distribution. This leads to marked seasonal variations in the availability and location of the vegetation and water needed for livestock production, which means that herds must be mobile and able to access scarce resources over extensive areas of land. In such non-equilibrium environments, mobile livestock production systems are a highly efficient use of natural resources (Scoones, 1994). Herds can be split according to the drought resistance of different livestock species. The most drought-resistant animal, camels, are herded in very remote locations when conditions are dry, and the arduous and hazardous task of camel herding is given to young men, who must survive on little more than camel's milk and wild foods for months on end. Decisions on where and when to move livestock are also influenced by security, relationships and agreements with other clans, access to markets and the need to minimize exposure to livestock diseases or disease vectors. Given the high levels of violent theft and destruction of assets in Somalia during the last 20 years, the mobility of livestock – which are easier to move than assets such as crops, supplies of grain or domestic items – could help explain why the livestock sector has suffered less than other economic sectors in some areas (Little, 2003).

Interventions supporting livestock and pastoralists

This section provides an overview of the main livestock and pastoralist support programmes implemented in southern Somalia since the mid-1980s.

Veterinary services and supplies before 1991

Until 1988 livestock development in Somalia followed a trend of underfunding and lack of attention from policy makers, similar to many other African countries. While in the 1980s livestock in Somalia accounted for 41.5 per cent of GDP, the budget for the Ministry of Livestock, Forestry and Range amounted to only 1 per cent of the national budget (Baumann, 1993). Structural adjustment in Somalia in the 1980s included attempts to reorganize state veterinary services, improve cost recovery and liberalize the importation and sale of veterinary pharmaceuticals. With falling operational budgets, veterinary services in already under-served pastoral areas declined even further and herders were forced to turn to black market supplies of veterinary medicine in rural markets. In 1988 the Somali government allowed private companies to import veterinary drugs for the first time and some drugs became available through private pharmacies. Pastoralists paid for veterinary medicines, as they had been doing for decades. In the colonial period herders paid for veterinary services and before the private importation of veterinary drugs in the late 1980s, herders paid for drugs either from government sources or informally in markets. Al Najim (1991) reports herders paying for veterinary drugs in the Bay region from 1976.

Post-1991 interventions: From relief to development

Free drugs and vaccinations

Civil war combined with drought in 1991–1992 caused an estimated 300,000 human deaths throughout Somalia and at least 50,000 losses in the Jubba Valley alone (Prendergast, 1997). From May 1991 to December 1992, the ICRC was the only major relief organization operating in southern and central Somalia. It began by providing food relief, health assistance, seeds and tools, veterinary support for pastoralists, and water and sanitation facilities. Activities were often carried out under extremely hazardous conditions and many Somali and three expatriate ICRC delegates were killed during the course of those operations.

The rationale for ICRC veterinary intervention was that the vaccination and treatment of livestock would increase health and therefore production, which would lead to improved livestock holdings and sales. ICRC seems to have done limited local consultation, but pastoralists would likely have identified animal health as a priority. When Save the Children UK was supporting a massive inflow of Somali returnees into the Ogaden in 1990, pastoralists told workers, 'If you can bring us nothing else, bring us medicines for our animals' (Holt and Lawrence, 1991). The ICRC project treated 2.5 million sheep and goats and 500,000 cattle and camels. Following the ICRC intervention other emergency relief livestock programmes were implemented by NGOs, funded by various donors. In 1993 the OFDA multi-sectoral programme for Somalia featured livestock assistance; the overall programme budget was $49 million. Among the UN agencies, both FAO and UNHCR conducted free or subsidized veterinary drug distributions in southern Somalia from 1992 to 1994. The Italian and Indian army veterinary corps of UNOSOM provided free veterinary treatments. During the post-conflict phase all livestock projects supplied subsidized or free animal health inputs.

Toward veterinary privatization and sustainability of services

As early as 1992 ICRC and CARE began to meet Somali veterinarians in order to discuss options for developing more sustainable veterinary services. These meetings related to ICRC's desire to initiate longer-term approaches to service delivery before their programme ended and CARE's institutional experience in small business development. The meetings indicated that support to private veterinary delivery systems was a logical option in an environment characterized by the absence of government and the spontaneous emergence of a whole range of private services, including health, education, electricity, water and communications. In this situation 'privatization' was something of a misnomer because there were no recognizable government services to reform or sell off to the private sector.

Following the onset of veterinary privatization initiatives by NGOs in northern Somalia, in early 1994 the 'veterinary privatization' theme was adopted by the European Commission of the European Union and from 1994 to 1996 an EC-funded programme coordinated 12 NGOs and the German bilateral agency GTZ in southern Somalia (Costagli, 1996). These organizations operated in areas not already covered by other NGOs in Sanaag and Sool regions in the north. The common approach of agencies within the EC programme was to provide support to Somali veterinarians and veterinary assistants in the form of business management training and kick-start credit packages. It was assumed that once veterinarians had established some kind of private business, they would then extend their services to pastoral areas at an unspecified later date (Costagli, 1996).

Although poor programme results and insecurity prompted most NGOs in the EU programme to close down work on animal health, the Italian NGOs UNA and Terra Nuova persisted, and with the agreement of their main donor, the EU, they developed an Itinerant Training Programme. The strategy was to improve the technical and business skills of Somali veterinarians with a view to making them more acceptable to livestock producers and traders as private practitioners. The programme also supported private groups of veterinarians in some regions, called 'veterinary associations'. Starting in 1997, the programme shifted its emphasis so that by its third phase in 2002–2003 it was training veterinarians on diagnostics and basic epidemiology. Using EU funds originally allocated to the Pan-African Rinderpest Campaign in Somalia, Terra Nuova vaccinated cattle against rinderpest and carried out rinderpest antibody tests in the Gedo region from 1998 to 2000.

In addition to the UNA and Terra Nuova programmes there were a range of other livestock activities. These activities were often a relatively small part of larger NGO programmes; it was difficult to compile an exhaustive list of them for this chapter because information was buried in general NGO reports, or not reported at all. The NGOs included those funded by emergency donors but also various religious and independently funded organizations. It was also difficult to determine what the religious organizations were doing and where, but it was thought that they provided some veterinary services free of charge.

With support from the EC Somalia Unit, FAO became more active in Somalia during this period and conducted comprehensive studies on livestock exports and marketing, including discussions with a wide range of stakeholders inside and outside Somalia (Stockton and Chema, 1995; EC Somalia Unit, 1996). In response to outbreaks of Rift Valley fever in the Saudi Arabian port of Jizan in 2000 and the consequent livestock import ban, FAO also supported technical assessments of Rift Valley fever in the region.

Supporting livestock trade and pastoralists livelihoods

Despite the conflict, a number of donor initiatives undertaken in Somalia could fall under development rather than relief. One reason for the development approach could be that the initiatives were part of broader regional initiatives and may have been designed originally for contexts quite different from conflict-ridden Somalia. In 2000, in order to complete the eradication of rinderpest in Africa, strengthen national animal disease surveillance systems and continue to encourage veterinary privatization, the EC developed PACE. It followed on from the Pan-African Rinderpest Campaign (PARC) and was implemented by AU/IBAR. The goal of PACE was to improve food security and reduce poverty, and its main strategies were capacity-building of AU/IBAR, the promotion of livestock trade from Africa by removing rinderpest, improving the control of other epizootic diseases and strengthening veterinary services. Although PACE was implemented in most countries by national government veterinary departments, the PACE Somalia programme, with a budget of €3.5 million, was implemented by NGOs and AU/IBAR.

Rinderpest in its classical form caused high mortality in cattle and therefore, was known to have a profound impact on the livelihoods on pastoralists in Africa (Catley et al, this volume). However, by 2002 it appeared that in southern Somalia, rinderpest was present in its lesser-known 'mild form' which caused very low mortality and minor clinical signs (Thomson, 2002). Consequently, when PACE Somalia began, rinderpest was probably not a priority for pastoralists or traders in southern Somalia. Pastoralists were not suffering losses due to rinderpest, and traders were moving large numbers of cattle to Garissa market in Kenya. In this situation, the eradication of rinderpest was of far more concern to the international community than to pastoralists in Somalia. During the four years of PACE Somalia, no measurable progress was made towards freedom from disease – the first stage of rinderpest eradication (Van't Klooster, personal communication, 2007).

The Community-based Animal Health and Participatory Epidemiology (CAPE) project was a DFID-funded project complementary to PACE. With US $8 million in funding over four years, the project covered pastoralist areas of the Horn of Africa region and worked with government to change policies and legislation to support privatized community-based animal health delivery systems. These systems involved working with communities to prioritize local livestock health problems and select pastoralists for training CAHWs, while medicines were supplied to the CAHWs through private vets who supervised their performance (CAHWs are described in more detail by Catley et al in this volume). Commitment to community-based approaches in Somalia stemmed from the long-term experience of practitioners and the positive impacts such approaches were seen to have.

Other programmes also focused on support to livestock marketing (particularly livestock exports) and primary-level veterinary service provision. The programmes assumed that the main constraints to the Somali livestock export trade were technical in nature and related to the sanitary requirements of importing countries. It was thought that improvements to certification systems and disease surveillance in Somalia would help restore exports to the Middle East following the trade bans, and would assist Somali traders to source new export markets.

Although the logic of enabling livestock exports from Africa through disease eradication and disease surveillance was initially supported by AU/IBAR, in 2003 its epidemiologists began to question the approach. Experience in Africa showed a widespread failure to eradicate epizootic livestock diseases relevant to trade, despite decades of aid investment (Thomson et al, 2004). Using the example of foot and mouth disease, it could be shown that meat from affected animals in Africa carried minimal risk in terms of disease transmission. It followed that, in the short to medium term commodity-based approaches were probably far more useful than disease eradication for promoting livestock trade from Africa.

These ideas caused some consternation in the OIE and with EC staff overseeing the PACE programme, because they questioned the rationale behind PACE and the disease focus and logic of international livestock standards. However, the Somali business community had already grasped the concept. Independently of technical advice from international livestock agencies or aid programmes, export abattoirs sprang up in Somalia to send chilled meat (which had a different set of sanitary requirements) to the Gulf. Between January and June 2006 abattoirs in Mogadishu and Beletweyne exported 169,946 head of livestock (FSAU Somalia, 2006b). For the cross-border trade of cattle to Kenya, it is likely that the major beneficiaries of the trade were wealthier livestock owners who were able to keep their herds mobile and take risks in times of hardship, in addition to herders living closest to the Kenyan border (Little, 2003). There is little evidence of how the trade affected women, herd structures, access to water or the enclosure of dry-season grazing, or on the access of poorer pastoralists to the markets.

Moving beyond animal health

From 1992 to 2002 most livestock interventions by the international community in Somalia were linked to animal health and support to live animal export trade. This began to change in 2003–2004 with two interventions. In 2004 the CAPE Project initiated private CAHW systems in southern Somalia. Called the Somali Communities Animal Health Project (SCAHP) and funded by the OFDA and DFID, this cross-border project covered border areas of eastern Ethiopia and the Bakool region of Somalia. From 2003 VSF Suisse, Cooperazione Internazionale (COOPI) and the Emergency Pastoralist Assistance Group (EPAG)[4] jointly supported privatized CAHW approaches in the drought-prone Gedo region. Funded by ECHO to respond to the 2002 drought in Gedo, the project, known as the Pastoral Assistance Programme, managed to break away from an almost exclusive focus on animal health. It included veterinary care, livestock production and nutrition, improved water access through renovation of non-borehole facilities, and livestock and marketing information. The programme was quite unlike the ICRC-led responses of 1992–1993. Despite Gedo having a reputation as being particularly unstable and difficult to operate in, it paid particular attention to impact and sustainability.

This change in strategies was supported by a number of studies and by a reorientation of the existing food security information systems, with increased attention to analyses. Oxfam Quebec was commissioned to do a livelihoods assessment for the AU/IBAR CAPE project prior to support to community-based projects, while a study by the War-Torn Societies Project on the livestock economy in Somaliland emphasized the disparity between international and community efforts. Starting in 2003 the FSAU began to provide more detailed livestock/pastoralist food security and early warning data to aid actors. This was as a result of the FSAU 2002 phase III mid-term review and a consultancy that examined how FSAU could improve information analysis and the dissemination of information on livestock (Simpkin, 2003).

Drawing heavily on experiences in southern Sudan (see Catley et al, this volume), three participatory impact assessments of CAHW interventions were conducted between 2002 and 2005, and to our knowledge, these were the only assessments of livestock interventions in southern Somalia since 1991 that examined the links between animal health and pastoral livelihoods. In the assessments of the VSF-Suisse/COOPI/EPAG project in Gedo region, Hopkins (2002; 2004) tracked the transition from a subsidized CAHW system through to a fully privatized approach using veterinary pharmacies. The second impact assessment conducted in April 2004 provided an excellent analysis of restraining forces, their effect on the project and lessons learned (Hopkins, 2004). The assessment report described the project's commitment to community participation, the impact on livestock health and production, the performance of private pharmacies and other indicators. Pastoralists perceived an increase in livestock numbers and a reduction in livestock mortality of up to 70 per cent for some diseases, due to easily accessible services. Livestock assets are a key measure of vulnerability, and the pastoralists assessed the project as hugely beneficial.

The CAHW project of AU/IBAR was assessed in 2005 and showed significant reduction in the livelihoods impact of disease treated by CAHWs compared with diseases not treated by CAHWs (Bekele, 2005). During an eight-month period, CAHWs treated more than 80,000 animals. Pastoralists attributed reduced livestock mortality and increases in milk, meat and income to CAHW activities. Increases in milk yields of 38 per cent and livestock weight of 10 per cent were recorded. Following the introduction of the CAHW service, the use of veterinary drugs sourced from black markets was also substantially reduced. The CAHWs participated in general disease reporting, an almost non-existent activity before the project, and 85 per cent of the total disease outbreak reports expected were submitted (Bekele, 2005). The participatory impact assessment demonstrated that CAHWs were viewed by pastoralists as highly accessible, available, affordable and trustworthy relative to other service providers, and they delivered a high-quality service.

Coordinating interventions under difficult circumstances

Coordinating from a distance

It is generally recognized that coordination plays a fundamental role in humanitarian response, although it is notoriously difficult to achieve (Donini, 1996; Minear, 2002). For livestock-related interventions, coordination was particularly important in southern Somalia because not only were relief and development programmes taking place simultaneously, but national rinderpest eradication required strong, central coordination with capacity to harmonize strategies in Somalia with those in neighbouring Ethiopia and Kenya. As a result of experiences under Operation Restore Hope and insecurity in southern Somalia, both general coordination and livestock-specific coordination were based in Nairobi, Kenya. In fact, all of the western NGO livestock programmes were run from offices in Nairobi.

Because coordination meetings took place in Nairobi, it was often impossible to include field staff or local Somali NGOs due to logistical issues and, in the case of Somalis, frequent problems with the Kenyan immigration authorities. Therefore, coordination meetings tended to be dominated by Nairobi-based expatriate staff and discussion was superficial because these staff simply relayed information from the field. Over time and as security worsened in southern Somalia, it became particularly difficult for expatriates to work in or visit certain areas to gain direct experience or observe activities.

In terms of livelihoods-based programming, this was a serious constraint because such programmes require high-quality, location-specific livelihoods analysis with communities and their representatives. There was obvious frustration among Somalis that programme decisions were being taken several hundred kilometres away, in another country. As one Somali elder in El-Ade, in the Gedo region said:

> If you think you can stay in Nairobi and decide for us here, you missed the point. Here we have the elders in control for security and management for the people. If you want to work with us you have to inform us about every step you take and we have to discuss it together. ... We are educated here and can decide for ourselves (Hopkins, 2002).

The Somali Aid Coordination Body

The two coordinating mechanisms for the livestock sector in Somalia were the PACE programme and the Livestock Working Group (LWG) within the SACB. The SACB was created by UNDP in 1994 with the aim of helping donors develop a common approach for allocating resources for Somalia. It was a multi-agency forum and its focus was development rather than relief; the United Nations Development Office for Somalia provided the secretariat. Although the SACB proposed broad stakeholder involvement, as described above there were problems in terms of the participation of Somalis living in Somalia. Similarly, in the absence of a central government, the SACB struggled to develop links with district or regional administrations.

The formation of the LWG by the SACB's Food Security and Rural Development Committee in 1998 was intended to provide a forum for in-depth technical discussions and improved coordination of livestock programmes. The forum was open to aid actors and comprised NGOs, donors, FAO, AU/IBAR and others. However, a crucial factor affecting the LWG and SACB was lack of any control over programme resources. This meant that coordination was largely sharing of information rather than concerted action on jointly agreed objectives. In 2001 the LWG started to become more active when a technical adviser in FAO Somalia became chair and the EC Somalia Unit started calling for an overall strategy for livestock development.

Two useful and tangible outputs were achieved by the LWG between 2001 and 2005. First, a code of conduct on community-based delivery systems and cost recovery in Somalia was drafted by AU/IBAR at the request of the LWG, which was endorsed by the LWG in 2004 after a year of discussion. Second, the EC Somalia Unit clarified the relationship between European development funds and ECHO (humanitarian funds) by re-stating the mandates of each department. These are modest achievements for the six-year lifespan of the LWG. However, no participating organization was responsible for its limitations; it was a product of the larger SACB structure and mode of operation.

Good coordination practices

Based on the results of an evaluation of aid coordination in Somalia conducted in 2004 (Wiles et al, 2004) and the research for this chapter, it is clear to us that:

- Coordination activities need benchmarks that ensure that coordination improves the livelihoods and food security of the primary beneficiaries.

The mechanisms for determining the benchmarks need to be built in from the start.

- A coordinating structure should be oriented toward beneficiaries rather than focusing on gaining credibility with donors, which means that coordination activities should be carried out in the field whenever possible.
- Sectoral and more general codes of conduct and strategies should be established early on in a democratic way, taking into account differences of opinion. The codes should be reviewed regularly and must be enforceable.
- Donors must agree ahead of time to utilize the coordination mechanisms. If there is competition for donor funds outside of the coordination mechanism, actors become less willing to share ideas and innovations within it.
- It is crucial that there be a dedicated, technically qualified and experienced coordinator for each sector. Without a coordinator, sectoral groups tend to lack focus, organization, innovation, problem-solving capacity and trust.
- Where the coordinating body is not the local authority there should be mechanisms to support the local authority, and there should be a clear handover strategy.
- Coordinating bodies should be able to support technical organizations in the areas of development and humanitarian practice.

Learning from the livestock interventions

The review of interventions above indicates a strong focus on veterinary programmes relative to other livestock-related interventions. The three main approaches were: strengthening the Somali livestock export trade, supporting private veterinary services by training Somali veterinarians and developing CAHW systems in pastoral areas. These could all be seen as development more than relief, despite the context of protracted crisis. To varying degrees, all three approaches were delivered by projects or programmes with overall objectives related to poverty reduction or improved food security.

However, it could be argued that the livestock owners in southern Somalia were worse off 13 years after the interventions began than they were at the outset, and there is some justification for questioning whether the strategies adopted by the international community in the livestock sector could have been improved. Certainly the Somalia experience provides the opportunity to learn some lessons for undertaking intervention in support of the livestock (and pastoralist) sector in a protracted crisis context. The following sections look at lessons on the importance of civil society support and community participation, the role of information in shaping responses and the importance of coordination.

Supporting Somali NGOs and community-based approaches: Missing the opportunity

What is striking about the majority of livestock initiatives described above is the lack of community participation, whether of traditional leadership, women, civil society organizations or local NGOs. When we asked agencies to comment on this, the common response was that it was 'too complicated' to work with communities in Somalia. Admittedly insecurity and conflict does make community involvement more difficult, time-consuming and inefficient, as it requires lengthy negotiations (Bradbury, 2003). However, Somali commentators have pointed out that the insecurity in southern and central Somalia in recent years has been manageable: armed conflict between the main militias and factions stabilized beginning in 1999, when the Rahanweyn Resistance Army consolidated control of the Bay and Bakol regions (Le Sage and Majid, 2002; Little, 2003). There are local administrations of varying commitment and capacity in all areas of south and central Somalia, and it is possible to work with many of them.

It is not only development theory that advocates community participation. Increasingly, humanitarian theory also advocates the involvement of communities as essential for effective humanitarian response (Smillie, 2001; Kathina Juma and Suhrke, 2002; African Humanitarian Action, 2004). According to Pingali et al (2005), intervention strategies in protracted crisis contexts will increase the resilience of food systems only if they include rebuilding local institutions and traditional support networks, reinforcing local knowledge, and building on people's ability to adapt and reorganize. The major constraints to using local communities adequately in the provision of emergency relief are short funding periods and the time required for negotiating equitable partnerships.

In protracted crises the arguments against local participation related to time pressures and short funding cycles are far less convincing. Even the development activities discussed here were weak in terms of community participation. For example, the disease surveillance activities of PACE Somalia were not based on building trust with pastoralists and linking surveillance to the provision of basic services. Apparently some livestock agencies regarded Somali communities as difficult to work with and a constraint to the achievement of agency objectives, rather than as partners to be involved in local analysis of problems and the design of solutions. These perceptions were captured in an evaluation of various projects (including livestock projects) funded by the Government of The Netherlands (see Table 7.4).

In some cases insecurity prevented international NGOs from establishing themselves in a given location. One solution to this problem would have been to utilize other organizations – such as indigenous NGOs – that could operate at higher levels of insecurity and had a more thorough understanding of local conflict and politics. In the case of Somalia, Islamic NGOs would have been the most appropriate. Islamic organizations[5] have grown rapidly in scope

Table 7.4 Contrasting perceptions of agencies and communities in Somalia

Characteristic	*Agencies*		*Somali community*	
	Self	*Somali community*	*Self*	*Agencies*
Dominant image	Benefactor	Aggressive	Coping	Imposing
Organizational characteristic	Structured	Anarchic	Negotiating	Coercive
Decision-making framework	Professional	Exploitative	Fate and Somali	Bureaucratic
Economic assumptions	Efficiency, effectiveness	Welfare recipients	Satisfying	Source of money
Negotiation assumptions	Neutral	Partisan	Decentralized	Authoritarian
Negotiation characteristics	Rational and objective	Unreasonable	Continuous	Regulation without dialogue

Source: O'Keefe et al (2001)

and importance in recent years and their dynamic expansion and success constitute an important trend in Somalia. Analysis of their location, successes and failures, mode of operation and objectives was carried out by Le Sage and Menkhaus (2004) and Novib (2003). While a minority of the organizations were perceived as pursuing radical Islamic agendas, mainstream organizations had had success in providing access to needed services (water, education, vocational training and health) and a popular vision of a political alternative to Somali clannism, violence and state collapse.

Despite the opportunities, there was very little contact between Somali NGOs and with the Nairobi-based SACB or donors. Indeed, the SACB LWG did not hold its first discussion on cooperation with Somali NGOs until December 2005. The meeting was relatively well attended by Somali NGO representatives (although it was unfortunate that the discussion mainly centred on whether an NGO run by Somalis but registered elsewhere was to be considered just another international NGO). They did not take into consideration that Somali NGOs from south and central Somalia did not have a mechanism for registration within Somalia, nor that they might have added value because of their Somali links (SACB, 2005). Furthermore, many of the veterinarians responsible for designing and managing programmes in southern Somalia seemed to lack understanding of rural development theory or practice. This deficit was reflected in the discussions at the LWG and the weaknesses of livestock programmes in terms of local capacity-building and working with Somali NGOs.

Information and learning

Evidence collected during this study indicates that programming decisions in the livestock and pastoralists sectors were taken by a restricted number of actors and based on limited information. Many assumptions were made and decision makers often missed the point in terms of livelihoods (see, for instance, the emphasis given to 'mild' rinderpest eradication in southern Somalia). These weaknesses can be attributed in particular to: (a) the poor linkages between food security/livelihoods analysis and the decision-making process; (b) the lack of effective impact-oriented project monitoring or assessment systems; (c) poor downward accountability; and (d) poor coordination. In Somalia food security and livelihoods information and analysis are more available than in other countries under protracted crises (such as DRC and Sudan). This is due to several initiatives and especially to the FSAU, which has been operating since 1994 with close technical links with a number of organizations and NGOs, including FEWS NET, SC–UK, CARE, UNICEF and more recently OCHA. (The FSAU has operated under FAO since 2000.)

The FSAU produces regular monthly food security and livelihoods reports, twice-yearly assessments, special research projects and livelihood baseline studies, plus more recently the Somali Livelihood Indicator Monitoring System, which provides market price data (cattle, camel, goat, sheep, milk). For the livestock sector these reports have improved markedly in recent years as FSAU developed a pastoralist livelihood monitoring system (Simpkin, 2003). There are still gaps in the FSAU livestock information and the baseline livelihood studies in southern and central Somalia are yet to be completed or updated. There are also missing variables such as meat export figures, and credible information on herd structure and off-take, although the process of obtaining this information has started. In future, the FSAU could provide information on the political economy of given areas, thereby assisting aid agencies working with local institutions to get a better understanding of clan dynamics. This would help them avoid fuelling local conflicts (see Little, this volume).

Despite the increasing quality of information available, up to late 2006 the link between information and response was not very evident. Although livestock agencies clearly recognized the importance of livestock and markets to pastoral livelihoods, programming did not progress beyond these fairly general associations. For example, few if any livestock interventions were based on a livelihoods analysis that disaggregated livestock assets by wealth or gender. Consequently, interventions were often targeted not at vulnerable pastoralists but at the livestock business as a whole (and in particular the export trade) or they were targeted at geographical areas. This problem is ably described by Hemrich (2005), and the FSAU is aware of the response shortcomings.

The FSAU encourages agencies to share analyses and engages in consensus-building; it thus favours interactive communication with decision makers. In late 2006 the FSAU added an 'Implications for Response' section to its *Food Security and Nutrition Update*. Some commentators felt uneasy about the section due to the implications if FSAU were to provide misleading programming advice to its readership. However, given the past domination of programming by a relatively small group of actors, the FSAU move towards provision of strategic advice is a welcome development.

At the response level, a series of well-coordinated livestock projects in southern Somalia implemented since the early 1990s might have generated a substantial body of knowledge on how to design and implement projects for the benefit of pastoralists. However, just like the long-term livestock interventions in southern Sudan (Catley et al, this volume), projects in southern Somalia concentrated on the measurement of process rather than impact. We can attain a sense of what was done and where in terms of project activities, but very little sense of the impact of these activities on people's food security or livelihoods. Thus repeated rounds of donor funds were allocated to interventions for which evidence of livelihoods impact was virtually non-existent. In this kind of environment, there appeared to be few incentives for NGOs involved in livestock work to think beyond animals and markets, and to examine impact on people. The participatory impact assessments of CAHW projects in southern Somalia described above represent the exception rather than the rule.

The overall impression gleaned from reviewing these experiences is that in a protracted crisis such as southern Somalia, characterized by lack of government, most agencies did as they wished, with limited accountability. Effective monitoring and more recently, real-time evaluation (ALNAP, 2003) are increasingly valued as ways to enable mid-course project revision. It is possible that more appropriate animal health inputs in southern Somalia could have been introduced much earlier if monitoring systems had encompassed downward accountability. This would have involved far closer consultation and participation of affected people (primary stakeholders) in the design and implementation of interventions, thereby reflecting genuine needs and priorities. In spite of vast expense it appears that the mass treatments and vaccinations against diseases, and years of training veterinarians, have not resulted in improvements in basic veterinary services for pastoralists.

The importance of actor coordination in protracted crises

'In my judgement, the continuing absence of effective coordination structures remains the soft underbelly of the humanitarian enterprise' (Minear, 2002).

Many analysts have called for UN agencies to be given authority to command players in the field during emergencies and recently, some donors are supporting the 'one flag' approach to coordination. Stephenson (2005) suggested that effective coordination includes strategic planning, information gathering and sharing, resource mobilization, common accountability frameworks, assuring a shared division of labour in the field, maintaining workable relations with host governments, vigorous leadership, and finally and most importantly, trust between organizations. Few of these characteristics of effective coordination were evident from our analysis of the coordination of livestock sector interventions in Somalia. Undoubtedly, limited coordination reduced the impact of interventions in terms of their support for the livelihoods and food security of livestock owners in southern and central Somalia.

Conclusions

Pastoralists have shown considerable capacity for adaptation to the crisis that has characterized southern Somalia over the last 15 years and some groups, such as livestock traders and large-scale herders, may even have benefited from the crisis. However, the livelihoods of most pastoralists have been seriously harmed, and vulnerable groups and poor households have been further marginalized. The responses of the international aid community have kept many people alive, but they have not contributed significantly to pastoral livelihoods.

The case study for this chapter showed that livelihoods-based approaches to improving food security were rarely applied for the livestock sector in southern Somalia. Agencies tended to use a superficial analysis of the role of livestock in livelihoods when justifying investments in traditional livestock programmes, despite an increasingly sophisticated body of livelihoods information emerging from FSAU. Interventions used conventional technical approaches focusing very much on veterinary matters: it was simply assumed that improvements in the sanitary arrangements for livestock trade or the further education of veterinarians would benefit vulnerable pastoralists. However, even after more than 10 years of implementation, there was very little evidence that these approaches resulted in livelihoods benefits for pastoralists in southern Somalia.

The study also noted that in the absence of recognized national and local authorities, community-based approaches could have provided important development opportunities. Unfortunately, international agencies and some NGOs (and in particular, advocates of the conventional rinderpest surveillance activities in southern Somalia) often argued that the area was a 'special case'. For these actors, the community-based approaches used successfully in other protracted crises could not be adapted for the southern Somali context. These views overlooked other long-term NGO experience in Somalia that demonstrated the capacity of communities to work with outsiders to organize themselves, and to design and fund local services.

We believe the evidence shows that higher-quality and more widely applied livelihoods-based livestock programming would be possible in southern Somalia through harmonization of donor strategies; improved understanding by donors, UN agencies and NGOs of livelihoods-based approaches; a single, empowered coordination body; and continued support for, and use of, the livelihoods analyses produced by FSAU. These changes would need to be accompanied by a core commitment to work with Somalis in Somalia as partners and as people who have a right to be involved in the design and assessment of aid programmes. This would require the involvement of agency staff who understood and respected Somali perspectives and did not dismiss the perspectives as irrelevant or wrong.

Livelihoods analysis shows that for Somali pastoralists, emergencies are expected and should not be regarded as entirely unpredictable. It follows that long-term programmes need to be developed that from the onset plan for the types of events that occur regularly in Somalia. Such a strategic shift should be guided by recent learning on post-conflict interventions, humanitarian actions, development in a complex emergency situations, real-time evaluation and cross-systems organization, and could be useful to many organizations currently needing to overcome institutional inertia in Somalia. A former UN Resident and Humanitarian Coordinator for Somalia has said:

> We need a paradigm shift that understands disasters and emergencies not as unfortunate occurrences that take place at the margins of human existence, but as reflections of the ways that human beings live their 'normal lives', and hence the ways that they structure their societies and allocate their resources. This paradigm shift will challenge some of the fundamental assumptions that underpin the humanitarian project as currently conceived' (Kent, 2004).

In 2001 it was proposed that a key strategy of the PACE programme should be widespread support to privatized CAHW systems in southern Somalia. These systems would provide direct benefits to pastoralists in the form of primary veterinary services, and help to build trust between communities and NGOs. Such trust would enable CAHWs to contribute to disease surveillance for rinderpest and other diseases, and would help NGOs to understand the broader livelihoods issues and lead to broader livelihoods programming. This approach was applied in a localized manner by some international NGOs. Impact assessments of these interventions were positive and such approaches need to be more widely applied.

References

African Humanitarian Action (2004) *Commitment to Collaboration: 'A New Partnership'*, background paper, International Symposium on Building Capacity and Resources of African NGOs, Addis Ababa, 5–7 December 2004, www.africahumanitarian.org/theme per cent203.pdf.

Al-Najim, M. N. (1991) 'Changes in the species composition of pastoral herds in Bay region', Pastoral Development Network Paper 31b, Overseas Development Institute, London.

ALNAP (2003) 'Humanitarian action: Improving monitoring to enhance accountability and learning', Overseas Development Institute, London.

Baumann, M. P. O. (1993) 'Animal health services in Somalia: Can centralized structures meet demand in the field?'. In M. P. O. Baumann, J. Janzen and H. J. Schwartz (eds) *Pastoral Production in Central Somalia*, pp. 299–321, Deutsche Gesellschaft für Technische Zusammenarbeit GmbH, Eschborn.

Bekele, G. (2005) 'Participatory impact assessment of the CAHW system: Bare district', Feinstein International Center, Tufts University, Boston, and African Union, Nairobi.

Bentley, C. (1989) 'Primary health care in northwest Somalia: a case study', *Social Science and Medicine* 28 (10): 1019–1030.

Bradbury, M. (2003) 'Living with statelessness: the Somali road to development', *Conflict, Security and Development* 3 (1): 7–25.

Bradbury, M., Menkhaus, K. and Marchal, R. (2001) *Somalia Human Development Report 2001*, United Nations Development Programme, Nairobi.

Catley, A. (1996) 'Pastoralists, paravets and privatisation: experiences in the Sanaag region of Somaliland', Pastoral Development Network Paper 39d, Overseas Development Institute, London.

Catley, A. (1999) 'The herd instinct: children and livestock in the Horn of Africa', Save the Children Working Paper 21, Save the Children UK, London.

Catley, A. and Walker, R. (1997) 'Somali ethnoveterinary medicine and private animal health services: Can old and new systems work together?', Proceedings of the International Conference on Ethnoveterinary Medicine, Research and Development, 4–6 November, 1997, Pune, India.

Costagli, R. (1996) 'Provision of veterinary services in conflict-stricken countries: The Somali case', MSc dissertation, University of Edinburgh, Edinburgh.

Donini, A. (1996) 'The policies of mercy: UN coordination in Afghanistan, Mozambique and Rwanda', Occasional Paper Number 22, The Thomas J. Watson, Jr Institute for International Studies, Brown University, Providence, Rhode Island.

EC Somalia Unit (1996) 'Improvement of livestock export marketing in the northeast region of Somalia. Report and recommendations', EC Somalia Unit, Nairobi.

Economist Intelligence Unit (2006) 'Somali country report, November 2006', The Economist Intelligence Unit, London.

FAO/World Bank/EU (2004) 'Somalia: Towards a livestock strategy. Final report', Food and Agriculture Organization of the United Nations, World Bank, European Union Report No. 04/001 IC–SOM, FAO, Nairobi.

FSAU Somalia (2006a) *Livelihood Baseline Profiles*, FSAU Somalia, Nairobi, www.fsausomali.org [Accessed November 2006].

FSAU Somalia (2006b) '2006 post-*gu* analysis', Technical Series Report Number V.9, 15 September 2006. FSAU Somalia, Nairobi.

Geshekter, C. L. (1993) 'Somali maritime history and regional sub-cultures: A neglected theme of the Somali crisis', presented at the First Conference of the European Association of Somali Studies, 6–10 December 1993, School of Oriental and African Studies, University of London, London.

Gorham, A. B. (1978) 'The design and management of pastoral development: The provision of education in pastoral areas', Pastoral Development Network Paper 6b, Overseas Development Institute, London.

Hemrich, G. (2005) 'Matching food security analysis to context: The experience of the Somalia Food Security Assessment Unit', *Disasters* 29 (S1): S67–S91.

Herren, U. J. (1990) 'The commercial sale of camel milk from pastoral herds in the Mogadishu hinterland, Somalia', Pastoral Development Network Paper 30a, Overseas Development Institute, London.

Holt, J. and Lawrence, M. (1991) 'An end to isolation: The report of the Ogaden needs assessment study 1991', Save the Children UK, London.

Hopkins, C. (2002) 'Emergency veterinary relief programme, Gedo Region, Somalia – Participatory impact assessment', VSF–Suisse, Nairobi.

Hopkins, C. (2004) 'Participatory impact assessment and evaluation of the pastoral assistance programme (PAP III), Gedo Region, Somalia', VSF–Suisse, Nairobi.

Janzen, J. (1986) 'Economic relations between Somalia and Saudia Arabia: Livestock exports, labor migration and the consequences for Somalia's development', *Northeast African Studies* 8 (2–3): 41–51.

Kathina Juma, M. and Suhrke, A. (eds) (2002) *Eroding Local Capacity: International Humanitarian Action in Africa*, Nordic Africa Institute, Uppsala.

Kent, R. (2004) 'Humanitarian futures: practical policy perspectives', HPN Network paper No. 46, April 2004, Overseas Development Institute, London.

Le Sage, A. and Majid, N. (2002) 'The livelihoods gap: Responding to the economic dynamics of vulnerability in Somalia', *Disasters* 26 (1): 10–27.

Le Sage, A. and Menkhaus, K. (2004) 'The rise of Islamic charities in Somalia: An assessment of impact and agenda', presentation to the 45th Annual International Studies Association Convention, 17–20 March 2004, Montreal.

Lewis, I. M. (1961) *A Pastoral Democracy*, Oxford University Press, Oxford.

Leyland, T. (1996) 'The case for a community-based approach with reference to southern Sudan'. In *The World Without Rinderpest*, FAO Animal Health and Production Paper 129, pp. 109–120, FAO, Rome.

Leyland T., Haji-Abdi A. O., Catley, A. and Hassan, H. S. (2006) 'Livestock, markets and food security in southern and central Somalia', a report to the Agricultural and Development Economics Division, FAO, Rome.

Lindley, A. (2005) 'Somalia country study. A part of the report on Informal Remittance Systems in Africa, Caribbean and Pacific (ACP) countries (Ref: RO2CS008)', ESRC Centre on Migration, Policy and Society, Oxford.

Little, P. D. (2003) *Somalia: Economy Without State*, James Currey Publishers, Oxford, and Indiana University Press, Bloomington.

Lybbert, T. J., Barrett, C. B., Desta, S. and Coppock, D. L. (2004) 'Stochastic wealth dynamics and risk management among a poor population', *The Economic Journal* 114: 750–777.

Mares, R. G. (1951) 'A note on the Somali method of vaccination against bovine pleuropneumonia', *Veterinary Record* 63 (9): 166.

Mares, R. G. (1954) 'Animal husbandry, animal industry and animal disease in the Somaliland Protectorate, Part II', *British Veterinary Journal* 110: 470–481.

Minear, L. (2002) *The Humanitarian Enterprise: Dilemmas and Discoveries*, Kumarian Press, Bloomfield, CT.

Nauheimer, H. (1993) 'Productivity of small ruminants in nomadic herds'. In M. P. O. Baumann, J. Janzen and H. J. Schwartz (eds) *Pastoral Production in Central Somalia*, pp. 183–199, Deutsche Gesellschaft für Technische Zusammenarbeit GmbH, Eschborn.

Novib (2003) *Mapping Somali Civil Society*, Oxfam Novib, Nairobi.

O'Keefe, P., Kliest, T., Kirkby, J. and Flikkema, W. (2001) 'Somalia: Towards evaluating the Netherlands' humanitarian assistance'. In A. Wood, R. Apthorpe and J. Borton (eds) *Evaluating International Humanitarian Assistance: Reflections from Practitioners*, pp. 19–38, Zed Books, New York and London.

Pearson, E. (1996) 'A review of the impact of civil strife on pastoralist education in Somalia. Gedo Region case study', Save the Children–UK, London.

Pingali, P., Alinovi, L. and Sutton, J. (2005) 'Food security in complex emergencies: Enhancing food system resilience', *Disasters* 29 (S1): S5–S24.

Prendergast, J. (1997) *Crisis Response: Humanitarian Band-Aids in Sudan and Somalia*, Pluto Press, London.

Reusse, E. (1982) 'Somalia's nomadic livestock economy: Its response to profitable export opportunity', *World Animal Review* 43: 2–11.

SACB (2005) 'Minutes of the December 2005 Livestock Working Group meeting', Somali Aid Coordinating Body, UNDP, Nairobi.

Sandford, S. and Habtu, Y. (2000) 'Emergency response interventions in pastoral areas of Ethiopia', UK Department for International Development, Addis Ababa.

Scoones, I. (ed.) (1994) *Living With Uncertainty: New Directions in Pastoral Development in Africa*, Intermediate Technology Publications, London.

Simpkin, P. (2003) 'Conceptual framework of a pastoral livelihood monitoring system for the Somalia pastoral livestock production system', FSAU Somalia, Nairobi.

Smillie, I. (ed.) (2001) *Patronage or Partnership: Local Capacity Building in Humanitarian Crisis*, Kumarian Press, Bloomfield.

Stephenson, M. (2005) 'Making humanitarian relief networks more effective: Operational coordination, trust and sense making', *Disasters* 29 (4): 337–350.

Stockton, D. and Chema, S. (1995) 'Somali livestock export market study', Joint EC–FAO Report, EC Somali Unit, Nairobi.

Thomson, G. (2002) 'Summary report on the Eastern African Regional Workshop on Mild Rinderpest', held in Nairobi, 17–19 June 2002, presented at the FAO-EMPRES Technical Consultation on the Global Rinderpest Eradication Programme, 30 September–2 October 2002, FAO, Rome.

Thomson, G. R., Tambi, E. N., Hargreaves, S. J., Leyland, T. J., Catley, A. P., Van't Klooster, G. G. M. and Penrith, M-L. (2004) 'International trade in livestock and livestock products: The need for a commodity-based approach', *The Veterinary Record* 155 (14): 429–433.

UNDP (2005) *Human Development Report, 2005*, United Nations Development Programme, New York.

Wiles, P., Hassan Farah, K. and Abdisalaam Bakard, A. (2004) 'Review of aid coordination for Somalia – Final Report October 2004', Somali Aid Coordinating Body, Nairobi.

PART III

Case Studies from the Democratic Republic of the Congo

CHAPTER 8

Crisis and food security profile: The Democratic Republic of the Congo

Koen Vlassenroot and Timothy Raeymaekers

A 'complex of war'

The Congolese wars (1996–1998; 1998–2003) constituted one of the most severe humanitarian disasters since World War II. They involved over six African nations and more than a dozen rebel groups. During the conflict, more than 3 million Congolese died either as a direct or an indirect consequence of armed confrontations (Coghlan et al, 2006). Many lost their physical and financial belongings, were displaced and suffered from the destruction of economic and social infrastructures.

At the root of this protracted crisis was the merger of local, national and regional conflict dynamics. From the early 1990s onwards, local competition for access to economic resources in eastern Congo was increasingly linked to a larger process of state collapse. Decades of economic mismanagement and patrimonial rule, the conversion of economic resources into political assets and profit-seeking activities by the ruling class caused a total collapse of the Congolese economy and prevented a process of formal institution-building. Informal networks of patronage monopolized their privileged access to economic resources, which were redistributed in order to secure political loyalty and prevent leadership challenges. When the resources needed to sustain this patronage system started to dwindle towards the end of the 1980s, President Mobutu was pushed by his Western allies to initiate a democratization process. Competition for political participation and for access to economic assets intensified and was increasingly based on ethnic criteria. In the eastern Kivu provinces, this democratization process had the effect of linking the exclusionary character of local land distribution to the issue of citizenship, which ultimately pitched entire ethnic communities against each other in a struggle over access to political and economic power.

In 1993 growing competition between indigenous and allochthonous communities led to a first outburst of violence, killing several thousand people. After the arrival of more than 1 million Rwandan Hutu refugees in 1994, these local dynamics of conflict became increasingly intermingled with a regional struggle for power in Africa's Great Lakes region. From refugee camps, Rwandan Hutu militias not only started attacking local Tutsi in Zaire but also challenged the new regime in Rwanda. This growing security threat resulted in the formation of a regional coalition, comprising Congolese communities of Rwandan descent, political opponents to Mobutu and neighbouring states Rwanda, Burundi and Uganda (soon to be followed by other countries in the region). Although initially aimed at dealing with this border security problem, this coalition had little difficulty in cutting right through the weakened Mobutu defence. Seven months later, it came to power in Kinshasa and installed Laurent-Désiré Kabila as the new president of the country.

What was hoped would be the end of a national and regional crisis, however, soon set in motion a process of political fragmentation. Growing disagreement between the new Congolese regime and its foreign supporters instigated the formation in 1998 of a rebel movement against the Kabila regime. This rebel movement, which again was supported by neighbouring countries Rwanda and Uganda, set in motion the formation of a regional war complex (including the involvement of several African nations), the institution of a multitude of local militias (leading to a total fragmentation of the politico-military landscape) and the dissolving of political agendas into local and individual interests. It was at this phase of the conflict that the Congolese war acquired its image of a struggle between criminalized politico-military networks for control over Congo's vast natural resources. Several reports have illustrated how the different belligerent parties oriented their struggle towards military control over mining sites, included several national armies and their Congolese proxies (United Nations Security Council, 2001).

From the start of the second conflict in 1998, the international community promoted peace talks between the different warring parties. In 1999, this resulted in the Lusaka Peace Agreement and the start of an Inter-Congolese Dialogue, to be followed by the deployment of a UN monitoring force. In July 2002, a peace agreement was signed between the DRC and Rwandan governments, leading to the withdrawal of Rwandan troops. In December 2002, the Congolese parties of the Inter-Congolese Dialogue signed an all-inclusive peace accord. After the approval of the final act of this peace agreement in 2003, a Transitional Government was put in place that was to create a new legal and institutional framework, prepare general elections and reform the security sector. In addition, the Mission of the United Nations in Congo (MONUC) was given a stronger mandate (under Chapter Seven of the United Nations Charter) and saw its military force expanded. Along with the International Monetary Fund (IMF) and the World Bank, the European Union and several of its member states significantly increased their assistance as well as their financial support of the government's transitional reform. The internationally supported initiatives to facilitate the peace process and promote regional stability included the Comité International d'Appui à la Transition (CIAT) (whose mandate ended at the official end of the transition period), the Multi-Country Demobilization and Reintegration Program (MDRP) and the Regional Conference for the African Great Lakes Region. This international commitment helped to establish some political stability and economic recovery and ensured the functioning of several transitional institutions. In addition, it facilitated the preparation of national elections (held in 2006) and redefined regional relations based on cooperation. As a result, the DRC now has a democratically elected parliament while its relations with Rwanda have improved considerably.

By early 2008 some caution remained, however, as this fragile political stability was regularly challenged by politico-military groups that refused to support the peace process. Rwandan Hutu rebels and autonomous Mayi-Mayi groups as well as rebels loyal to the dissident Laurent Nkunda continue to be a considerable source of insecurity in the Kivu provinces (eastern DRC). At the same time, security reform stumbled due to lack of confidence between the warring parties and lack of willingness to address the key issues. Large-scale corruption was omnipresent and systematic human rights abuses were being reported with increasing frequency. In rural areas, structural violence, the imposition of forced taxes, summary detentions and systematic sexual violence still conditioned people's daily lives. Congo's elections were unlikely to lead to democracy if not accompanied by a new political and governance culture. This new culture was far from being realized; the peace process and electoral campaign pointed at one of the crucial weaknesses of the current political process. Rather than paving the way towards democratic and transparent rule, they seemed to reconfirm patrimonial rule and the use of public positions for private gain.

Violent conflict and local food systems

In most parts of the DRC, the food security situation worsened considerably following the outbreak of violence in 1996. While already before the war the Congolese population was faced with increasing food insecurity and generalized poverty (mainly as a result of decades of economic mismanagement and patrimonial rule), since 1996 remaining productive capacities and economic opportunities were further reduced and reached the lowest levels in the world. In 2002, about 80 per cent of the population lived below the poverty line of $0.20 per day (AfDB/OECD, 2006). Other social indicators showed similar realities. The infant mortality rate rose from 125 per 1,000 live births in 1990 to 170 in 2000, and maternal mortality from 800 to 2,000 per 100,000. While life expectancy in Africa in 2002 was 51 years, in the DRC it was only 42 years (AfDB/OECD, 2005). In the eastern DRC, more than 90 per cent of the rural population had no easy access to safe drinking water. There were considerable regional disparities in poverty figures, however. The average national annual income per capita in real terms in 1998 was estimated at $110 (FAO, 2000), but in the eastern areas of the country it was only $32 (AfDB/OECD, 2005).

The same disparities existed in the food security situation. Even if reliable information remained very scarce, available data indicated that particularly in the eastern parts of the DRC, the nutritional position of large parts of the population had reached dramatic levels. Chronic infant malnutrition was over

45 per cent for children under 5 years old (the national average was 38.2 per cent (UNICEF, 2001)), and daily diets were deficient in micronutrients, proteins and lipids. In contrast to the eastern parts of the DRC, in the capital Kinshasa, important supply changes, innovations and coping strategies (such as the development of peri-urban and urban agriculture) as a result of the war had facilitated a continuing food supply and allowed most Kinois to deal with the most severe effects of chronic food insecurity (Tollens, 2005).

There were many causes of food insecurity in eastern DRC. Generalized insecurity (especially in rural areas) led to a decrease in production, limited access to markets and reduced financial means. In South Kivu, according to statistics of the Inspection Provinciale de l'Agriculture, Pêche et Elevage (Ipapel), the amount of cropland cultivated dropped by 29 per cent overall between 1996 and 2004 and by more than 50 per cent in the most remote areas. General food production dropped by 12 per cent, vegetable production by 42 per cent and the production of cereals by 33 per cent. In North Kivu, livestock activity shifted from cattle raising (the number of cattle was reduced by more than 50 per cent during the war) to small livestock activities. The main constraints to production were limited access to land and tools, pillaging of harvests and animal stocks by armed groups, and lack of treatment of diseases of small stocks and plants. Lack of access to markets and growing isolation of remote areas had further discouraged production of food. In addition, the collapse of the road network and the presence of armed elements reduced access to markets and the transport of food from production areas to centres of consumption. The presence of military checkpoints along the main roads also limited access to markets. Following the formal end of the war in 2003, illegal taxation by regular army units at checkpoints acquired a systematic character. In some areas, food traders were losing more than 20 per cent of their goods during their trip from the production site to the market, which reduced their income considerably and led to increased food prices at local markets.

Another critical constraint to food security was a growing lack of financial assets. Daily income per inhabitant decreased from $1.31 in 1973 to $0.23 in 2000 (AfDB, 2005). One survey concluded that in Walungu (South Kivu) during 2004, 72 per cent of households had a monthly income of less than $30 and 22 per cent had an income between $30 and $50 (Bahirire, 2004). Another study based on a survey of 840 households in South Kivu revealed considerable shifts in the number of meals per day. While before the war about 60 per cent of the population ate three meals a day, during the war the figure was only 3.3 per cent and in 2004 about 11 per cent. During the war, 83.3 per cent ate only one meal per day, but by 2004 the figure had decreased to 45 per cent (Diobass, 2005). A survey in Kamituga (South Kivu) revealed that sales of agricultural products by individual farmers in 2004 had dropped by roughly a quarter in comparison with the pre-war period (1995–1996) (Vlassenroot and Raeymaekers, 2004).

Under Mobutu, attempts of civil servants to exploit the population (as part of the dominant fend-for-oneself culture promoted by the political centre) became part of the daily living conditions. One strategy was to impose illegal taxes and opportunistic arrangements on traders, which had the effect of reducing the profit margins of trading activities. In addition, these taxes limited the financial resources and disturbed the access and distribution of food. In Bandundu and Equateur for example, these taxes paralysed riverine commerce of agricultural produce, which had an impact on the food security in Kinshasa because basic foodstuffs had to be imported, making them prohibitively expensive for the capital's indigent masses (Rackley, 2006). The reduction in farmers' income was also caused by a shortfall of economic exchange between urban centres and insecure rural areas, and by the importation of basic necessities that were formerly produced locally. Before the war, Kinshasa's main provider of beans, maize and meat were the Kivus, but after the war it was cheaper to import agricultural products from Zambia, Tanzania, South Africa and other countries.

The coping mechanisms adopted by the most vulnerable households to deal with the effects of declining income further affected food security. These strategies included the reduction of quality and quantity of meals, a preference for crops that have a minimal risk but low alimentary value, exchange of labour for food, and displacement (Save the Children, 2003). Migration to mining sites, which became a very popular coping strategy in resource-rich areas, caused neglect of local food production and a considerable increase in food imports and prices.

A final factor in reduced food security was the disappearance of local structures (including agronomists) to assist and guide local farmers. The lack of government assistance resulted in decreased production and capacities to deal with diseases such as cassava disease and tracheomycosis (coffee wilt disease). Recent surveys revealed that less than 15 per cent of rural households had access to knowledge and advice from agronomists, while assistance had become crucial due to a sharp increase in disease. In the Kivus, more than 75 per cent of cassava fields were infected by disease, which had reduced production by 30 to 60 per cent, and in the most highly affected areas by as much as 80 per cent. Production of bananas (an important staple food but also an important source of income), was also affected by disease and insects and in some regions declined by 50 per cent. The shortfalls in production consolidated the market dependence of many farmers, which in certain regions was already in place before the war, for guaranteeing their access to food. A lack of financial means, however, forced the rural population increasingly to consume cheap but less nutritious products, which caused a shift in local food patterns.

Key actors and features of response

The limited governance capacity of the fragile Congolese state was echoed in food security policies. Pre-war government structures at provincial and local levels virtually disappeared or lost most of their financial and logistical means. In regions where rebel movements took control, rebel structures were not able to revive agricultural structures and facilitate the delivery of basic food needs through effective institutions. Even more serious was the fact that during the war rebel movements used access to vulnerable people as a political tool and in several cases impeded relief operations from reaching communities in territories under government or militia control.

Even before the start of the conflict, numerous international agencies tried to deal with the effects of such state failure on the food security situation. After 1996, food security interventions further intensified and concentrated foremost on physical assistance to refugees, IDPs and the most vulnerable and food-insecure people. These interventions went beyond the provision of food to include the provision of shelter and material equipment to ensure minimal agricultural activity. A key stakeholder for providing and distributing food aid was the WFP. During the first years of the DRC crisis, WFP activities were undertaken as an emergency operation (EMOP). Between 2000 and 2003, WFP mobilized about 260,000 metric tonnes of food commodities, targeting mainly IDPs in eastern parts of the country. In January 2004, the EMOP was replaced by a two-year PRRO, which aimed at providing food assistance to war-affected people and comprised the three basic components of relief, refugees and recovery. In the most war-affected regions, however, the implementation of the relief component took priority. Another key player in the distribution of food was ECHO, which was the DRC's largest donor of humanitarian assistance, with an allocation of more than 180 million euros between 2000 and 2005. As part of its emergency assistance, ECHO created and supported a number of nutrition centres in eastern DRC and established supplementary feeding centres and therapeutic centres based at local hospitals or health centres.

Attempts were made all along to shift assistance from emergency relief to rehabilitation interventions and to integrate humanitarian interventions into a larger strategy of peace-building. The delivery of food aid and the support to nutrition centres were accompanied increasingly by rehabilitation efforts, initiatives such as food for work and cash for work, and support of farming and fishing activities. In addition, some agencies adopted longer-term programmes and integrated strategic plans including protracted relief and recovery operations. Even if development activities remained very small in scale, in several regions the international donors' willingness to assist medium- to long-term programmes resulted in at least a modest economic recovery and the re-establishment of security in inaccessible areas. One example was the food for work (FFW) and cash for work (CFW) approach, which was adopted by a number of international agencies as part of their programmes to rehabilitate the local road structure. In general, the rehabilitated and accessible roads tended to improve security conditions and had a positive impact on food security. Surveys in Masisi (North Kivu) revealed that economic exchanges and access to income increased, displaced populations returned and survival mechanisms were reinforced after the rehabilitation of the Sake–Masisi road. In the same region, agricultural production increased by 50 per cent and the number of livestock by 400 to 500 per cent. FFW and CFW initiatives were also linked to demobilization and reintegration efforts of ex-combatants.

However, most food security interventions suffered from important constraints. Many international agencies complained about a lack of information flows. Basic data about the food security situation of populations living in more remote or conflict-prone areas often were not available because local information systems were absent or had collapsed. In addition to data on the evolution of food prices, only limited information on the nutritional status of populations existed. WFP conducted vulnerability assessments, Save the Children carried out a number of household economy analyses and a livelihoods study focusing on land, Asrames organized some local assessments and World Vision International and Médecins Sans Frontières did some small-scale surveys on the food security situation in some parts of eastern DRC. Few efforts were made by aid agencies, though, to fill this information gap. Most assessments and analyses conducted by aid agencies focused only on specific issues (such as humanitarian conditions, IDP movements, food needs, etc.) about which information was needed for short-term, supply-based interventions. Little attention was given to the wider and structural causes of food insecurity, the structural shifts in food systems, local mechanisms and responses to food shortages, social transformations and the longer-term impacts of food security interventions.

Another complaint was weak or lacking coordination between humanitarian actors. With over a million displaced people in the eastern parts of the country, humanitarian intervention required robust coordination on the part of national and international development agencies. This weakness was partly connected to the lack of protection in more remote areas, which prevented international agencies from operating in the most severely affected areas. Nevertheless, a number of mechanisms were established that aimed at coordinating humanitarian activities. The main objective of these initiatives was to increase the efficiency of humanitarian interventions and the level of understanding of the local humanitarian situation and needs. At the provincial level, OCHA centralized information on humanitarian issues and interventions and invited agencies to weekly meetings to exchange information and discuss the security situation. OCHA's coordination mechanism also aimed at facilitating the communication between local political/administrative authorities and aid agencies, strengthening the coordination of interventions and defining priorities and zones of intervention through the organization of regular seminars. From the start of the war, international agencies, their local partners and local authorities set up a series of commissions aimed at facilitating the response to urgent humanitarian needs. These commissions became more effective with increased support from UN agencies, which then coordinated them. One of these commissions was the Food Security Commission, headed by the provincial inspectorate of agriculture and facilitated by the Emergency Operations Service of the Food and Agriculture Organization of the United Nations. However, its activities were limited mainly to the exchange of information and did not include the definition and preparation of (joint) strategies of intervention.

Despite efforts to adopt longer-term programmes, food security interventions in eastern DRC continued to be planned with a short-term perspective (Levine and Chastre, 2004). An assessment of ECHO interventions demonstrated that although the distribution of food aid and support of nutrition centres had achieved their goals (even when demands and needs were often greater than the available quantities and quality), large-scale food aid imports involved the risk of slowing down agricultural recovery (Michael et al, 2004). Most agencies failed to adequately consider the purchase of local food, even if local purchase can serve to stabilize markets and stimulate production. No coordinated local food purchase programmes were developed to bridge the gaps in the food pipeline. Finally, most programmes were not directed at reinforcing livelihoods.

Access to markets and land

Analysis of the food security situation in eastern DRC indicates that livelihoods are determined to a large degree by the institutional context. Institutions constitute the critical factors in facilitating or blocking access to assets and thus are crucial to the food security position of households. In complex political emergencies such as the DRC, violence not only influences the available resources, options and choices of households, but also has an impact on the existing institutional context, or on the policies, institutions and/or processes through which households negotiate the use of assets and the configuration of livelihoods strategies. However, on the rare occasions when food security interventions took the institutional context into account, it was in a very marginal way.

One structural factor that had a preponderant effect on food systems during the Congolese conflict was access to land. The mediating structures that regulate that access were a critical factor for inhibiting or facilitating livelihood capabilities and choices. In the case of eastern DRC, access to land was the main driver of local conflict and evolved into an important asset for warring parties as the conflict proceeded. In the most critical areas, even before the outbreak of violence in eastern DRC, local farmers' access to land was reduced as a result of governmental land tenure policies aimed at introducing entrepreneurial agriculture and at redirecting profits derived from land ownership to the advantage of a class of largeholders. This process turned many farmers who had been small landowners into simple labourers and increasingly eroded traditional coping mechanisms. The effects of the capitalization of land were felt most in the densely populated areas of eastern DRC such as Lubero and Masisi, where crop patterns shifted from nutritious crops to low-risk but protein-deficient crops such as manioc. The reduction in size of cultivated plots was one of the underlying structural causes of famine as early as 1989 (Pottier and Fairhead, 1991).

The war amplified the effects of limited access to land. Land became one of the main dynamics of conflict in Ituri, Masisi and other regions, and warring factions used their control over land to reinforce their political position to the disadvantage of small farmers. Chapter 10 analyses these land-related structural factors of poverty and food insecurity. It argues that in eastern DRC alternative strategies developed by households to address reduced access to land and to guarantee their food security eventually induced a number of shifts in the organization of land access and food consumption patterns. These changes attracted only limited attention during food security interventions, in spite of the fact that food security can be addressed in a number of ways through interventions that focus on land access, such as strengthening the legal position of rural populations and strengthening the role of community representatives in land issues.

Another institutional variable that conditions people's access to food is access to markets. The economic crisis that affected Zaire during the 1970s and 1980s led to mass impoverishment as well as the economic marginalization of peasant households. The extremely clientelistic organization of rural markets during this period not only favoured the accumulation of capital in the hands of an emerging capitalist elite, it also contributed to the emergence of a series of commercial middlemen who, profiting from the economic isolation of rural households, succeeded in reaping the benefits from rural and transborder trade. The Congolese wars largely reconfirmed this exploitative relationship between rural capitalists and peasant households. Lack of infrastructure, insecurity and large price fluctuations led to growing disparities between rural and urban markets, which simultaneously strengthened the position of commercial middlemen and urban retail traders. To this trend should be added the opportunistic behaviour of neighbouring countries, which profited from Congo's economic disorganization to exploit local markets (an example is the coffee trade from North Kivu to Uganda). Peasant households continued to be dependent on the access to economic markets to substitute their incomes curtailed by the war. Hampered by the exploitative organization of rural markets, peasants often had no alternative other than to withdraw into subsistence agriculture. At the same time, however, the growing pressure on land property continued to push the same peasants back to the market, where they were again exploited by local and regional entrepreneurs. Chapter 9 uses the case of Beni-Lubero to discuss the problems that lack of market access poses to the diversification of livelihoods.

Against this background, most food security programmes still tended to focus on food availability rather than access to food and markets. Cash-based interventions such as microcredit initiatives began only at a late stage to form part of some aid programmes, most of which provided only basic food aid. An important challenge for food security programmes was – and is – to become more knowledgeable about, and responsive, to the role institutions play.

Timeline of conflict in DRC

1990 Mobutu announces end of the single-party system and institutes democratization process.

1991 Riots by unpaid soldiers in Kinshasa and other cities.

1993 Political confusion and creation of rival pro- and anti-Mobutu governments and parliaments. Inter-ethnic violence in Katanga and North Kivu Provinces.

1994 Genocide in neighbouring Rwanda kills up to a million Tutsis and moderate Hutus. Over a million Hutus, including members of the Interhamwe *genocidaires*, flee to Zaire for refuge. Rwandan Hutu militias use Zaire as a base and start campaign against the new regime in Rwanda and Tutsi minorities in eastern Zaire.

1996 The Alliance of Democratic Forces for the Liberation of Congo-Zaire – led by Laurent Kabila and backed by Rwanda, Uganda, Burundi, Angola and Zimbabwe – starts military campaign against Mobutu.

1997 Rebels topple Mobutu. Laurent Kabila becomes new president and renames Zaire the Democratic Republic of the Congo.

1998 Kabila expels Rwandan and Ugandan troops from Congolese territory, which in return back new rebel movement against Kabila. Zimbabwe, Namibia and Angola support Kabila. Start of the second Congolese war.

1999 War leads to military stalemate due to growing fragmentation and internal quarreling between Ugandan- and Rwandan-supported rebels in Lusaka. The Lusaka Peace Accords are signed by the governments of DRC, Angola, Namibia, Rwanda, Uganda and Zimbabwe and major rebel forces, calling for a cessation of hostilities by all forces in DRC. Fighting continues between rebels, the government, foreign armies and numerous foreign and domestic militias. The UN Security Council authorizes establishment of MONUC to assist in implementing the cease-fire.

2000 Fighting between Rwandan and Ugandan troops in Kisangani; anti-Kabila alliance further disintegrates.

2001 Kabila assassinated by his bodyguard and replaced by his son, Joseph.

2002 In July a peace agreement is signed between DRC and Rwanda that commits Rwanda to withdrawing troops; in September, Uganda also agrees to withdraw its troops. In December, following almost a year of negotiations, the government and main rebel groups sign a peace agreement in South Africa that paves the way for the institution of a transitional government composed of main warring factions and opposition members.

2003 French-led EU operation *Artemis* instituted to end atrocities in Bunia (Ituri), followed by strengthening of MONUC's mandate and operational capacities. Kabila signs a transitional constitution; a transitional government and parliament are put into place. Security conditions in eastern Congo remain precarious.

2004 Mutiny in Bukavu (eastern DRC) of Banyamulenge and Tutsi forces and fighting in North Kivu between the Congolese army and renegade soldiers from a former pro-Rwanda rebel group. Unrest in Kinshasa as coup plot fails.

2005 Transitional parliament adopts new constitution. The Independent Electoral Commission registers Congolese voters and organizes referendum that approves the post-transition constitution. In Ituri, nine UN peacekeepers are killed by militia. The security situation is also very volatile elsewhere in eastern DRC.

2006 The constitution is signed into law by President Joseph Kabila, who wins the presidential elections after defeating Jean-Pierre Bemba in the second round. National and provincial parliaments are elected in the first democratic elections in four decades. In December 2006, the newly elected president is inaugurated. In North Kivu, fighting between the Congolese army and renegade soldiers led by Laurent Nkunda causes massive displacement.

2007 The provincial parliaments elect new provincial governors and appoint members of the Senate. A peace deal is signed between Laurent Nkunda and the Kinshasa government.

References

AfDB (African Development Bank) (2005) *République Démocratique du Congo. Document de stratégie par pays axe sur les resultants 2005--2007*, AfDB, Abidjan.

AfDB/OECD (Organisation for Economic Co-operation and Development) (2006) 'Country studies: Democratic Republic of Congo', *African Economic Outlook 2005/2006*, 16 May, www.oecd.org/dataoecd/13/8/36746740.pdf [accessed 31 January 2007].

Bahirire, C. (2004) [untitled], mimeo.

Coghlan, B., Brennan, R., Ngoy, P., Dofara, D., Otto, B., Clements, M. and Stewart, T. (2006) 'Mortality in the Democratic Republic of Congo: A nationwide survey', *Lancet*, 367 (9504): 44–51.

Diobass (2005) [untitled study], Diobass, Bakavu.

FAO (Food and Agriculture Organization of the United Nations) (2000) 'Global information and early warning system on food and agriculture' in FAO (Special Report), *Crop and food supply situation in Kinshasa and the provinces of Bas-Congo and Bandundu of the DRC*, FAO, Rome.

Levine, S. and Chastre, C. (2004) *Missing the Point. An Analysis of Food Security Interventions in the Great Lakes*, HPN Network Paper No. 47, Overseas Development Institute, London.

Michael, M., Thouvenot, V., Hoerz, T. and Rossi, L. (2004) *Evaluation of ECHO's Actions in the Democratic Republic of Congo*, Prolog Consult, Brussels.

Pottier, J. and Fairhead, J. (1991) 'Post-famine recovery in highland Bwisha, Zaire: 1984 in its context', *Africa*, 61 (4): 437–470.
Rackley, E. (2006) 'Democratic Republic of the Congo: Undoing government by predation', *Disasters*, 30 (4): 417–432.
Save the Children (2003) *Household Economy Analysis of the Rural Population of South-western Bwito, Rutshuru, North Kivu*, Save the Children UK, London.
Tollens, E. (2005) 'Food security in Kinshasa: Coping with adversity'. In T. Trefon, (ed.), *Reinventing Order in the Congo – How People Respond to State Failure in Kinshasa,* Zed Books, London, pp. 47–64.
UNICEF (United Nations Children's Fund) (2001) *Multiple Indicator Cluster Surveys (MICS2)*, www.childinfo.org [accessed 20 January 2007].
United Nations Security Council (2001) *Report of the Panel of Experts on the Illegal Exploitation of Natural Resources and Other Forms of Wealth of the Democratic Republic of the Congo*, document S/2001/357 (12 April).
Vlassenroot, K. and Raeymaekers, T. (2004) *Conflict and Social Transformation in Eastern DR Congo*, Academia Press, Ghent, Belgium.

CHAPTER 9

Conflict and food security in Beni-Lubero: Back to the future?

Timothy Raeymaekers

Abstract

This chapter looks at the extent to which household coping strategies in the region of Beni-Lubero in the DRC contribute to an effective adaptation of livelihoods, and the extent to which food security interventions have facilitated or blocked the process of developing resilience. Beni-Lubero was chosen because of its characteristic as a 'neither war nor peace' situation. Indeed, in this context, households often show a great degree of inventiveness in their short-term coping strategies. Whether or not this inventiveness leads to greater resilience is the main question of this chapter. Finally, recommendations are given for adapting existing food security programmes in complex and volatile contexts.

Introduction[1]

During conflict situations households often show a great degree of inventiveness in the face of food crises (Flores et al, 2005). However, it is not clear whether this inventiveness always leads to their livelihoods becoming more resilient. This chapter looks at the extent to which household coping strategies in the region of Beni-Lubero in the DRC contribute to an effective adaptation of livelihoods, and the extent to which food security interventions have facilitated or blocked the process of developing resilience. Beni-Lubero was chosen because of its characteristic as a 'neither war nor peace' situation (Richards, 2005; see also Vlassenroot et al, 2006); while some parts of the region can still be defined as acute emergencies, others are characterized by more 'stable' communities that have adopted more or less 'effective' responses to the crisis (which may or may not result in greater resilience to food insecurity). This contrast between the different sub-areas of Beni-Lubero might be explained in part by the more structural dimensions of livelihoods vulnerability during protracted crises – which crystallize around the problem of *institutions*.[2]

'Coping strategies' is a term commonly used in the literature to refer to short-term household responses for dealing with external stress and shocks. These might include, for example, a temporary change in household diets or temporary/circular migration. 'Adaptive strategies' refer to the way households respond to adverse events over the medium to long term. Ideally, such adaptation should lead to greater resilience of households towards external shocks. However, as will become clear, the fact that households attempt to make their livelihoods less vulnerable to protracted food crises does not necessarily make them more resilient.

Answers to questions about livelihood adaptiveness will be sought primarily in how particular coping strategies are linked to wider livelihood opportunities. As suggested in the literature on social capital, and elsewhere, a key determinant of grassroots development – particularly the development of livelihoods – is the existence of autonomous, or extra-communal, ties that link households to wider economic and social opportunities. For a community or household to develop livelihood resilience, there must arise 'a need and inclination to reach out beyond the original spatial, economic and mental boundaries of the group' and supplement basic strategies of self-reliance with more centrifugal tendencies that form bridges with other groups and opportunities (Woolcock, 1998). In this regard the role of institutions is particularly important – or more specifically the mediating structures that define and facilitate livelihood *access* to certain *assets*. Institutions are important here as the 'agencies and conventions' that inhibit or facilitate household capabilities and choices (Ellis, 2000).

This chapter focuses on access to land and markets as two important regulating structures that determine the development of livelihood opportunities. Other factors, such as the decline of basic infrastructure and the absence of a responsive framework of social service delivery, also play determining roles in this process. The progressive collapse of government institutions in the DRC during the 1980s and 1990s was not only a crucial determinant of the civil war; it continues to play a part in denying large parts of the population access to vital resources (Vlassenroot and Raeymaekers, 2004a; Vlassenroot et al, 2006). The focus on land and market access mechanisms is particularly relevant in this regard because it highlights the *inventiveness* of households in ensuring access to livelihoods assets during protracted food security crises – while simultaneously serving as a warning against overly optimistic (functionalist) accounts that largely deny the underlying *structural* dimensions of poverty and vulnerability.

A key point of departure for this chapter is that subsistence farming on its own is not a sufficient strategy for households that must provide for themselves when their livelihoods are threatened.[3] Households must depend instead on a diverse portfolio of activities and income sources, which include the networks, organizations and institutions that enable such diversity to be secured and sustained. Therefore, part of this chapter is dedicated to the issue of livelihoods diversification – i.e. of how households have tried to diversify their activities and income sources in response to the current food crisis. In the second part of the chapter, these attempts at diversification are weighed against the networks, organizations and institutions that determine how diversification is being sustained or blocked. Finally, recommendations are formulated with regard to food security interventions in Beni-Lubero, with particular attention to the problem of 'linkage' and institutions.

Research methods

The method used as a basis for analysis in the chapter was a combination of basic data collection and participative workshops. The first part consisted of the elaboration of a baseline study on food security interventions, based foremost on interviews with key organizations. For the second part of the study (on livelihood diversification and adaptation), a series of workshops was organized in nine localities (Masereka and Luhoto in central Lubero; Musasa and Muhangi in western Lubero; Mighobwe, Kikuvo, Kirumba, Kayna and Kanyabayonga in southern Lubero). Participants were asked to answer three questions: (1) Which factors negatively influence your agricultural production/revenue, and what degree of importance do you assign to each?; (2) What strategies do you develop both in the short and medium term to confront these negative influences on your production/revenue?; and (3) What has been the impact of food security interventions? Each exercise was divided into two parts, one of which focused on agricultural production and the other on agricultural revenue. A series of preponderant factors mentioned by participating farmers were clustered and the most important identified. Those identified as having most influence on agricultural production were climate, environment, demography, access problems and institutional factors. Those most influencing agricultural revenue were price, quantity put on sale, organization of the market, 'taxes', means of transport, storage and conservation and transformation.[4]

An important dimension of this research method was that influencing factors were never treated in isolation, but rather as part of a whole, as part of a cluster so to say; the aim was not to estimate the absolute relevance of one or the other factor, but to evaluate how households reacted to a particular combination of factors. A general group discussion at the end of the workshops aimed at evaluating the relationship between the factors that were mentioned as most important and the interventions carried out in the food security domain.

Introduction to Beni and Lubero

The territories of Beni and Lubero are situated in the province of North Kivu. Because they constitute the largest part of the province, they are commonly called '*le grand nord*' (while '*le petit nord*' comprises the smaller territories of Walikale and Masisi). North Kivu has three agro-ecological zones, which can be divided into low, middle and high altitude areas, each with a diversity of micro-climates and cultures. There are two major farming seasons, one from mid-September to mid-January and the other from March to July. Agricultural calendars vary according to zone; in central Lubero, the sowing and weeding seasons extend from January to March and from July to September; the harvest seasons are from April to June and from October to November, with potatoes harvested in June and December. The principal crops produced for consumption in central Lubero are potatoes and grain; those produced mainly for sale are onions and potatoes. In the rest of North Kivu, the main crops produced for consumption are manioc, potatoes and beans, while those produced for sale are manioc, beans, rice ('paddy' or peeled) and palm oil.

The three food economy zones in Beni-Lubero are:

1. The highland area, east of the road from Lubero to Kanyabayonga, with its economic centres in Masereka and Luhotu (*'chefferie des Baswagha'*). This region traditionally functions as the agricultural backyard of North Kivu. Vegetables such as cabbages and onions are transported to Kisangani and Kinshasa.
2. The plateaus centred around Butembo, which produce mainly subsistence crops such as bananas and manioc. Butembo also functions as a commercial centre for the entire eastern part of the DRC, with links to Kisangani and Kinshasa.
3. The lowland areas located both in western Lubero and the Graben (or Central African Rift Valley). While the western lowlands consist primarily of tropical forest, the Graben is covered mainly by Virunga National Park. The area of Manguredjipa (in the far eastern part of the province) also hosts some minor mining activity, mostly for coltan and other tin derivates.

The three regional trade zones are:

1. Kasindi: Situated on the Uganda–Congo border, this market constitutes the 'gate to the East' of the Kivus. Agricultural products and minerals from eastern Congo province are traded for household products and prime necessities from East Africa and Asia. Through Kasindi, Nande traders maintain links with such distant places as Thailand, China and the Philippines.
2. Butembo: Forming the bridge between the eastern and western markets, it provides both the interior (Oriental Province, Equateur) and Kinshasa with products from the Kivus and beyond.
3. Kirumba–Kayna: Provides the link between the fertile highlands of Lubero and the southern markets of Goma, Bukavu and Katanga.

Economic conditions

A main characteristic of Beni-Lubero is its dynamic commercial activity. Although its precise origins are still subject to debate, three main factors have contributed to this economic dynamism: strong bonds of ethnic kinship among the inhabiting Nande community;[5] a favourable economic and geographical climate; and the proximity of country borders. Since pre-colonial times, border cities like Aru (on the Congolese–Sudanese border) and Kasindi (on the Ugandan border) have been lucrative trading centres for agricultural and mineral products from the provinces of North Kivu and Oriental. The rich soil and temperate climate of northern North Kivu are also favourable for the cultivation of crops such as cabbages and maize, which are in demand in other parts of the country. The expansion of cattle ranching in Lubero has turned the province into an important exporter of beef to Kisangani and the capital Kinshasa. These trade opportunities provided enormous possibilities of accumulating wealth in both the 'formal' and the 'informal' economies, which gradually formed the basis for the emergence of a Nande commercial class (MacGaffey, 1987).

During the mid-1970s and 1980s, two developments strengthened this commercial dynamism: land reform and liberalization of the mining sector. As in the rest of the Kivus, the Bakajika land law favoured the large-scale appropriation of vacant land by a new class of rural capitalists (Mafikiri, 1994). These concessions were very often used not for production purposes, but as a basis for capital accumulation. During the mid-1980s in Beni-Lubero, Nande entrepreneurs started to use their contacts with (local) government officials to acquire large swaths of territory in their home regions. (In 1994, three quarters of all liberated land titles in Luhotu (Lubero) belonged to Nande entrepreneurs.[6]) Rather than reinvesting the profits from land ownership in agriculture, the Nande used the income to support development of commerce; retributions from land sales and tenant farmers were largely invested in transport, trade and construction.[7] At the same time, small commerce proved to be an attractive extra-agricultural supplement for small-scale farmers who were confronted increasingly with large-scale appropriations. Contrary to Bwisha (Masisi) for example, where farmers still remained overwhelmingly dependent on food *production* during this period (Pottier and Fairhead, 1991), Lubero and Beni experienced a gradual shift towards more 'capitalist' forms of economic production such as cash crops (for example, coffee) and petty commerce: in the early 1990s, over a quarter of local farmers supplemented their agricultural incomes with small-scale commercial activities, principally the sale of agricultural and household products (Mafikiri, 1994). This segment should be added to a large group of landless farmers that worked on the entrepreneurs' plots to produce cash crops destined for the market. In Lubero during this period, a low-value, small-distance trade rapidly expanded that operated in tandem to the long-distance bulk trade that exchanged foodstuffs from North Kivu to Kinshasa and Kisangani. On the one hand, Nande businessmen increasingly started to control the agricultural output of places like Masisi and Lubero, where foodstuffs were being produced for the urban centres. On the other hand, small farmers in Lubero could benefit from price fluctuations between medium and small markets: like in Masisi, profits could be made in Lubero and Beni by peasants (especially women) who carried foodstuffs to and from places where the big lorries could not pass. Because of the expanding gold trade through Nande territory in this same period,[8] bigger traders could also combat devaluation by using this 'dependable hard currency' (Vwakyanakazi, 1987) instead of dollars or the (rapidly devaluating)

Zaire.[9] This engagement of both small-scale farmers and urban businessmen in the so-called informal economy at that time is locally explained as a central element in overcoming the country's economic crisis for the Nande. Finally, it also contributed to a gradual emancipation of female peasants in the Lubero highlands, who now assumed better positions in their households thanks to informal commerce and their participation in private credit institutions such as rotating credit groups or *likelemba* (Mafikiri, 1994).[10] This evolution would nonetheless be seriously disrupted with the advent of the first and second Congo wars.

From civil war to 'neither war nor peace'

While a detailed description of the Congolese wars (1996–1997 and 1998–2003) lies outside the scope of this chapter, some elements have had particularly important effects on Beni-Lubero's food security situation. On the one hand, the Congolese conflicts set in motion an intensified fragmentation process, the effects of which were felt across the entire political and economic landscape. The subsequent division of the Kivus into several military strongholds led rebels and local businessmen in Beni-Lubero to make informal pacts that had important consequences for the wider political economy. On the other hand, systematic pillaging and looting led to severe losses of economic assets, which seriously undermined local food security.

Starting in 1999, the territories of Beni and Lubero were occupied by the Rassemblement Congolais pour la Démocratie-Mouvement de Liberation (RCD-ML), founded in 1998 as a split-off from the original RCD movement.[11] While the RCD was supported by Rwanda, the RCD-ML acquired the support of Uganda. As a result, the territories of the *grand nord* quickly developed into a separate military stronghold, in opposition to surrounding rebel groups. From 1999 to 2003, Beni-Lubero was attacked by the Mouvement pour la Liberation du Congo (MLC) and RCD-National[12] (from the west), the Union de Patriotes Congolais (UPC)[13] (from the north) and by the RCD (from the south). The Nande's economic position was challenged increasingly by economic actors from neighbouring Ituri, mainly northern Hema traders who acquired the support of the UPC. Nande businessmen were even murdered in Bunia and other places during the UPC's military campaign in Ituri. Meanwhile, competing RCD-Goma elements obstructed traditional Nande trade with the rest of the Kivus and Katanga by blocking the road from Lubero to Goma. To cope with this loss of market opportunities, traders from Beni and Butembo increasingly entered into 'prefinancing' relationships with the occupying rebel movement RCD-ML, in which political patronage was exchanged for reduced tax duties on import–export trade. But while this relationship guaranteed these traders a certain 'protection' from the rebel movement, it could not make up for the dramatic loss of opportunities and assets caused by the ongoing crisis.[14]

During the entire Inter-Congolese Dialogue, the leadership of the RCD-ML made concerted efforts to win a place in the new Transitional Government, which was finally inaugurated in June 2003. Mbusa Nyamwisi, the leader of the RCD-ML, became Minister of Regional Cooperation, which it was hoped would stimulate economic collaboration between Central Africa's former adversaries (in particular Uganda, Rwanda and Burundi). Abbé Malu Malu, the former dean of Butembo University, ran the first post-war Congolese election process as the head of the national Electoral Commission. The Congolese transition process offered certain Nande leaders the possibility of re-establishing their links with Kinshasa. Business people from Butembo took the opportunity to regain lost properties in Kinshasa, and even to sign new contracts.

In the meantime, the territories of Beni and Lubero evolved from civil war to a situation that could best be described as 'neither peace nor war' (Richards, 2005). Beginning in December 2004, the region was relatively calm following the installation of mixed Congolese army units (primarily along the Lubero–Rutshuru border). The Congolese army and MONUC also began to carry out regular missions to search and dislodge Forces Démocratiques pour la Libération du Rwanda (FDLR) (Rwandan Hutu rebels) from the North Kivu countryside. Nonetheless, insecurity still reigned beyond the main axes. Southern Lubero in particular continued to be an emergency area, hosting nearly 40 per cent of North Kivu's IDPs.[15] Another region that continued to suffer from insecurity was western Lubero, home to the rural Mayi-Mayi militias and some dispersed FDLR units. The relative 'peace' that had come to the urban areas was coupled with a smouldering, low-intensity conflict beyond the urban borders.

The impact of war

The particular situation that characterized Beni-Lubero during the years from about 1998 to 2005 (a transition from military stronghold to a situation of 'neither war nor peace') had serious consequences for its population. In terms of food security, short-term access to food was influenced foremost by the factors of destruction of infrastructure, disruption of regional trade and plummeting levels of agricultural output (including small and big livestock). Longer-term access to food was hampered mainly by structural impediments, some of which are discussed below.

Infrastructure

Beni-Lubero's infrastructure has suffered immensely from the pillaging and destruction of passing armies and militias. One consequence was the serious deterioration of the local road network. As a result, farmers had to invest much of their energy in maintaining road tracks, which deteriorated into muddy pathways, especially during the rainy season. The regular displacement of households led to the destruction of local habitats, which were rebuilt only gradually due to a lack of skills and equipment. The rudimentary dwellings that peasants constructed were regularly occupied by the military. As a result, many peasants preferred to construct provisional homes for fear of destruction by the military and for lack of financial means to acquire more durable materials (plastic sheeting, for instance, cost about $15 on the local market). Health services became less functional due to lack of medical supplies and to insufficient and unqualified technical staff. This was particularly the case in zones like Manguredjipa, where as of 2005 only 7 out of 17 structures were registered in the provincial drugs register. Since the collapse of the Zairian state system, teachers were not paid and educational opportunities decreased drastically due to the incapacity of parents to pay school fees and due to the destruction of infrastructure. The level of primary and secondary education dropped to minimal levels, especially in areas with difficult access. Schools seldom had teaching equipment, and often lacked even the chairs and tables necessary to host their pupils.

The war also affected industrial infrastructure. The factories of Beni and Butembo were an important military target (although not plundered as heavily as South Kivu). The immediate result of the war was that the modest industrialization process that slowly took root during the 1990s came to an abrupt halt. Factories like ENRA (timber), SAIBU (soap) and SOBEKI (beverages) came to operate on a fraction of their pre-war capacity. As a result, basic necessities like bottled water and soap had to be imported from neighbouring countries. The complete disintegration of the banking sector made it almost impossible for local entrepreneurs to rebuild that lost infrastructure, let alone to call for foreign investment.

Regional trade

Obviously, the decline of infrastructural capacity had a pernicious effect on regional trade. On the one hand, both manufacturers and traders were confronted with steep price increases for the imported products that gradually took over the local market. On the other hand, the drop in overall national demand caused a spiralling downward trend in local agricultural prices. In 2003, a sack of beans (100 kg) sold for $12 (in the countryside) and $28 (in Beni), as opposed to $80 during the pre-war period (Jackson, 2003). In Beni-Lubero local production of coffee (traditionally the main export product) decreased from over 30,000 tonnes in 1990–1991 to less than 6,500 tonnes in 2001–2002. The drop in local production led to a partial reorganization of the market: while in 1990–1991 Beni was still buying over 70 per cent of local production, the balance shifted to small coffee buyers in Butembo, who eventually took around 60 per cent of local produce.

Prices for agricultural products varied from one entity and one season to the other. In Pinga (Walikale), 1 kg of manioc flour cost five times more than in Oicha (Beni). Prices also varied according to the different stages of production, commonly designated the *vache maigre* (the period of sowing and weeding) and the *vache grasse* (the period of harvesting); during the *vache grasse* prices tended to be substantially lower due to an increased availability of foodstuffs. Price variations could be explained in part by the differing levels of security, which influenced households' access to their fields and stocks. In the mining areas, prices for foodstuffs tended to be substantially higher because they were imported from outside the area by road or airplane. Due to these decreasing transport facilities, as well as growing insecurity, these price fluctuations benefited mostly the large bulk traders operating from the urban centres, contrary to the small farmers-traders that became increasingly tied to the market.

Agriculture

Local agriculture was yet another victim of the war. As a result of insecurity and the repeated displacement of households, local productivity fell to minimal levels. In North Kivu in general, the productivity of beans fell 72 per cent, of manioc 53 per cent and of bananas 45 per cent (OCHA, 2005). In Lubero, overall agricultural production dropped 67 per cent; this situation was slightly more positive than in Masisi, where in 2001 manioc production fell by 96 per cent and bean production by 91 per cent (since 2001, however, the situation has improved considerably; see Vlassenroot, this volume). Other causes affecting agricultural production were the repeated plundering of harvests by armed forces and the inaccessibility of certain parts of Lubero (in particular the axis Butembo–Manguredjipa and the axis Musasa–Muhangi).

The situation was similar for livestock production: due to repeated pillaging by rural militias, traditional herding grounds in southern and central Lubero lost almost all their livestock during the war. (Cattle production began a modest recovery in 2005.) An increasing number of households invested in the breeding of small livestock such as goats, chicken and cobay (a kind of guinea pig introduced as an animal protein by several NGOs). By 2005 fish farming was also gaining popularity around Oicha (in northern Beni).

The loss of economic assets did not necessarily lead to a change in the local division of labour. The principal activities in Beni-Lubero remained agriculture and cattle raising (about half of overall household activity) and petty trade (roughly one-third of overall household activity). This relationship was also reflected in the sources of revenue, which were mainly the sale of agricultural produce (roughly one-third of total household revenue) and petty trade (between 20 and 35 per cent). An important difference with the rest of North Kivu was that more of the food in the food basket was bought than produced: households bought roughly half of their food at local markets, while subsistence agriculture represented only one-third of the basket, and food aid barely 1 per cent.[16] This confirms our hypothesis that it is not subsistence agriculture as such, but rather the linking of livelihood assets to wider socio-economic opportunities, that constitutes the main chance for households to emerge from the food security crisis.

Structural impediments to market access

Structural factors continued to affect the access of households to wider opportunities, particularly those involving (sub)regional markets. Although border towns like Kasindi and Aru (in Ituri) still functioned as competitive markets for food (such as manioc, palm oil or beans) and cash crops (such as coffee and papaya), transborder trade in the area had not led to a favourable trade balance for producing peasants. At the time of our research, the trade ratio between the rural and urban areas was 9 to 1 – which meant that the countryside supplied nine times more than it received. Notwithstanding the inventiveness of households towards the food crisis, and the theoretical presence of large-scale opportunities (in the form of competitive markets), structural factors still impeded households from equitable access to labour opportunities and markets.

The situation in the region of Mangina (territory of Beni) provided one clear illustration of such structural impediments. For a number of years, transborder traders in Mangina encouraged farmers to plant cash crops like vanilla, cocoa and papaya, as a substitute for the withering coffee production (which suffered highly from the disease *tracomycose*). The arrangement between farmers and traders usually evolved as follows: an agreement was reached on the price of the specific crop, which was usually accompanied by a pre-payment to the farmer for expected produce. To ensure that his deal was respected, the trader then engaged another farmer – or sometimes even his own 'security' personnel – to check on the evolution of the prefinanced produce. This control mechanism was intended in addition to prevent the farmer from selling his crops twice (for example to a competing trader who might offer more for the same produce). The control mechanism ended up tying the farmer twice to the prefinancing trader: first in terms of production (which was partly paid for in advance) and second because the farmer was obliged to sell his produce to the same oligopolist trader. This production–trading mechanism serves as a prime example of the continuing structural exploitation of peasants by Beni-Lubero's urban bulk traders, in spite of the market opportunities for the peasants' products that theoretically would open up.

'Post-conflict' indicators

Although technically a post-conflict area, by early 2007, Beni-Lubero could still be defined in part as an emergency situation, particularly in the south. It had been relatively calm following the installation of mixed Forces Armées de la République Démocratique du Congo (FARDC) units[17] along the region's southern border in December 2004. Accompanied by an Indian MONUC battalion, these units were meant to supervise the respect for the cease-fire agreement as reached with dissident army soldiers following armed clashes at the end of 2004. In addition, the Congolese army, together with MONUC, carried out regular missions to search and dislocate FDLR (Rwandan Hutu rebels) that continued to roam the Kivu countryside. While this mixed FARDC–MONUC presence secured (more or less) the main centres (including the main road from Lubero to Kanyabayonga), the interior still suffered from insecurity. The far south of Lubero continued to be a fragile area, particularly after the resurgence of armed clashes in its direct vicinity during the winter of 2006.[18] Additional pressure came from an already extremely high population density (up to 253 inhabitants/km^2 in the highland areas), a strong drop in agricultural production and very limited access to crucial resources such as land and water.

The determining factors of this 'post-conflict' emergency situation continued to be infrastructure and food (in)security. Accessible roads were still basically absent, especially away from the urban centres, and farmers continued to have to invest much of their energy in maintaining their access to fields and the market. The destruction of housing and the obstacles to reconstruction, and the dire lack of health and education services, contributed to the emergency conditions, although health indicators in western and southern Lubero improved slightly beginning in early 2005, following the extension of security into those areas (author interviews with medical workers in the area).

Household responses

The previous overview presented the direct consequences of the conflict on aspects affecting food security and livelihoods in Beni-Lubero. In turn, the deterioration of assets stimulated households to adopt a number of short-term and longer-term strategies to cope with the impact of food insecurity. Short-term strategies are usually referred to as *coping* strategies, and include 'temporary responses to declining food entitlements [that] are characteristic of structurally secure livelihood systems' (Davies, 1996). *Adaptation* strategies in contrast are longer term, and defined as the process of 'changes to livelihoods which either enhance existing security and wealth or try to reduce vulnerability and poverty', including the 'diversification' of assets and income sources in anticipation of future shocks (Davies and Houssain, 1997).[19] Ideally, such *adaptation* strategies should make households less vulnerable (more resilient) to external shocks, but not all paths of adaptation are successful. Households might adapt their livelihood portfolios in such a way that they become more, instead of less, vulnerable to future shocks.

One coping strategy adopted in the region of Beni-Lubero, for example, was the adaptation of existing consumption patterns, including temporary changes in diets and a reduction in the number of rations per day. Before the war, households managed to maintain a diverse diet, including regular consumption of vegetables and proteins. In response to increasing pressure on livelihood assets, Beni-Lubero's households gradually shifted their diets away from protein-rich foods; fish and meat virtually disappeared. Following the war, manioc constituted the most important – and sometimes the only – staple food. The poorest households (up to 50 per cent in inaccessible zones like Manguredjipa and Mbau) did not even have regular access to oil, salt and soap. Due to the war, the number of meals had fallen from three to one for most poor households. In many cases, children were the only ones to eat in the morning, and the only true meal was eaten in the evening by the woman/household head, who had spent all day working in the field. Acute malnutrition rates were high, varying from 5 to 10 per cent in the province of North Kivu as a whole, and reaching 18 per cent in inaccessible areas. Chronic child malnutrition was around 60 per cent for the entire province of North Kivu (WFP, 2004).

Consumption patterns differed slightly from region to region, however. Lubero households responded to the crisis primarily by reducing the number of meals per day, while that strategy figured as a last resort in Beni territory. The consumption of seeds and genitors was a first strategy in Beni, and only a last resort in Lubero (see Table 9.1 for the results of a survey by WFP).

Regarding longer-term *adaptation* strategies, the food security crisis in North Kivu led to various attempts to diversify sources of the household food basket. One of these strategies was the involvement in petty trade. As indicated above, this activity traditionally forms an important potential *adaptation* strategy for households to vulnerability and shocks. Although it is usually assumed that households during crisis situations mostly withdraw into subsistence, this was only partly true for North Kivu, where subsistence farming contributed only about 50 per cent of household income.[20] The existence of subsistence stocks and markets, therefore, potentially offers an important opportunity for households to reduce their vulnerability to shocks.[21] The greater diversification of livelihoods in Beni, for example, could be explained in part by its more favourable position in relation to commercial markets. Beni's vicinity to important border markets such as Kasindi meant that local households could sell substantially more of their production than could households from more isolated regions (OCHA, 2005). Compared with neighbouring regions, Beni also produced more cash crops such as vanilla and cocoa, which could be exported to regional markets.

As in the 1980s, the favourable position of Beni's and Lubero's households in relation to agricultural markets did not necessarily translate

Table 9.1 Responses to crisis

Territory	*Strategy*	*Coping Strategies Index*
Lubero	1. Reduce the number of meals per day	5.7
	2. Harvest premature crops	5.5
	3. Reduce the portions per meal	5.4
	4. Consume seeds and/or genitors	4.9
Beni	1. Consume seeds and/or genitors	9.8
	2. Harvest premature crops	7.9
	3. Consume less appealing or cheaper food	5.4
	4. Reduce the portions per meal	5.3

Source: WFP (2004)

into greater livelihood resilience, however. OCHA even noted a higher degree of vulnerability of households in Beni in comparison with the rest of North Kivu. This trend could theoretically be explained by two factors. The first factor was the continuing high dependence of Beni and Lubero's households on food *production* which, as said, dropped tremendously because of constant insecurity, loss of agricultural know-how and so on. A second, more important factor, however, consisted of Beni-Lubero's symbiotic relation with the market. In comparison with the rest of North Kivu, the household food basket in this region depended much more on the selling and buying of food: households sold roughly half of their produce and bought two-thirds of their food on local/border markets. Again, this demonstrates that more structural factors were at play that impeded peasants from benefiting from their commercial opportunities. Contrary to the idea that peasants largely withdraw into subsistence farming during food crises (Hyden, 1999), Beni-Lubero's households continued to use agricultural markets to supply their food basket. Nevertheless, their potential profit was immediately creamed off by other, more powerful actors that operated in the sphere of political and economic 'mediation'. These obstacles to livelihood resilience are discussed in the next paragraphs.

Obstacles to wider economic opportunities

In order for a household confronted with severe food crises to survive, its livelihood must depend on a variety of sources and activities. But for livelihood diversification to be a successful adaptation strategy, households must have access to wider economic opportunities. One such economic opportunity ought to be provided by the market as an institution, and another by access to land.

Access to markets

An important factor often overlooked in the analysis of farmers' access to markets is the role played by local 'mediating structures' (Ellis, 2000). During conflict these 'structures' usually extend to include additional armed groups, which come to both constitute and replace historical clientelist institutions such as customary land regulation and commercial intermediaries. As a consequence of the Congolese conflict, transaction costs associated with market exchange rose considerably due to insecurity, wrecked rural infrastructure, proliferation of taxes and the disappearance of credit systems. On the local market, these transaction costs emerged mainly in the form of taxes, which subsequently enlarged the price wedge between consumer prices and producer prices. Farmers in central and southern Lubero, for example, had to pay a plethora of taxes and confront tributes and harassment before reaching the market. Taxes included 'hygiene' taxes, city taxes, military taxes and so on, for which farmers never received services or public goods in return;

they were mostly pocketed by privatized state agents or the military, who had no connection whatsoever to the state treasury. This form of illicit taxation was by no means new: Fairhead (1992) and Newbury (1984) for example, talk about the organized robbery of North Kivu's peasants by a coalition of state and traditional administrators, army and police from colonial times until far into the 1980s. What was different during the Congo war was the sheer randomness of taxation by rebels and militias, and the fact that these taxes were not any more levied ad valorem (i.e. on the basis of the value of sold produce) but instead depended on the number of times farmers went to market (regardless of the amount of produce they put on sale). The result was that such taxes took away roughly *half* of the farmers' gross revenue from the sale of agricultural products.[22]

This arbitrary taxation contributed to making vulnerability increasingly widespread. Apart from the somewhat 'official' taxes mentioned above, farmers were regularly confronted with various forms of harassment on the way to market, including roadblocks and theft. In emergency areas like southern Lubero, for example, households had to pay a ratio of roughly 1.5 kg of manioc a week to Rwandan Hutu militias, who had their operating base on the Lubero–Rutshuru border. This ratio did not cover the 'military' tax, which was to be paid to the Congolese Army for every market entry.

The organization of Beni-Lubero's markets had an even greater effect on agricultural income. Lacking information and contacts, farmers in Beni-Lubero had to deal with various intermediaries, who continued to monopolize farmers' access to these markets. Similar situations have been reported in comparable situations of protracted conflict; they work mostly to the benefit of traders who 'are willing to take the risk involved' in operating under such security constraints (Longley and Maxwell, 2003; see also Collier, 2003). The intermediary phenomenon was particularly widespread in Beni-Lubero, which could be explained in part by the pervasiveness of petty trade in the area.

In the absence of systematic data, two examples indicate the extent of the problem of intermediaries. The first example is that of the trade in manioc flour. In Beni-Lubero, this semi-finished product is often used as an alternative during the period of the '*vache grasse*', i.e. when overproduction leads to a a drop in manioc prices (manioc is the most important staple food). Instead of the usual $5 offered per bag (100 kg) of raw produce during this period, manioc flour is sold at $11 per bag by southern and central Lubero's farmers. The net revenue gained from the seasonal production of manioc in the form of manioc flour can thus be estimated at roughly $25–30. However, this positive calculation acquires an entirely different dimension when compared to the gross priced earned by the intermediary trader: $14 for the same bag of manioc flour. Other things being equal, the percentage farmers lose to this form of intermediary trade roughly equals 50 per cent of their estimated net revenue. A similar dynamic exists in the coffee trade. Commercial intermediaries often sell farmers' produce at the border market in Kasindi, where they receive five times the average gross price offered to small farmers ($1.50 instead of $0.30).

In order to circumvent this loss of their agricultural earnings, farmers in Beni and Lubero developed a number of strategies to reach the market more directly. To profit from their produce, they combined direct selling practices with regular visits to larger markets. Different markets were often held on the same day, so farmers had to choose the most favourable. The most important criteria for deciding were: distance (i.e. time spent selling agricultural produce); price (certain products could be bought at more favourable prices at specialized markets); harassment (some markets were renowned for their bothersome tax agents); and availability of basic necessities for sale.

Table 9.2, shows that many strategies adopted by Beni and Lubero's farmers with regard to the market were mostly 'reactive', i.e. they consisted mostly of trying to avoid overtaxation and a negative influence of price fluctuations on their selling and buying operations. To avoid being taxed for every market entry for example, women sometimes sent small children to the market with part of the produce; administrators are usually reluctant to tax children when they enter the market. Some peasants (especially women) also bring goods to the market together, so they can divide the tax burden. A lot of speculation also occurs in order to anticipate certain price fluctuations, but due to the massive impact of bulk trading these speculations mostly turn out negatively for the peasants. To sideline this influence, in western Lubero, some peasants even set up small selling points (called *boutiques*) in their doorways to sell to passers-by instead of going through monopolistic intermediaries. Apart from

Table 9.2 Strategies for the sale of agricultural products

Price	*Taxes*	*Market organization*	*Conservation*
• Speculation • Cheating (for example use of measuring cups instead of weights) • Going to 'specialized markets' • Brewing of maize into aracque, an alcoholic beverage • Grinding of manioc/maize into flour • Long-term barter with other farmers • Informing oneself on the exchange rates in Butembo • Selling potatoes intended for seed	• Collective marketing of products • Sending minors to the market (less noticeable) • Payment in kind • Selling directly from the field • Negotiation with tax collectors • Developing social relationships with tax collectors • Benefiting from tax collectors' lack of cash	• Informing oneself on the profitability to the intermediaries of the products • Defining the margin of profit to be left with the intermediaries • Setting up of individual selling points	*Maize* • oiling the grains with palm oil or ashes *Potatoes* • drying on shelves • harvesting carefully (to avoid rotting due to damage) *Onions* • awaiting full maturity • storing in a dry place

Source: Author compilation

these methods, farmers had very few instruments at hand to circumvent the bleeding of their profits by intermediaries and taxing agents.

The underlying explanation for the decidedly disadvantageous position of farmers in relation to the market may be found in the lack of linkages of small farmers to networks that went beyond their immediate bonds of solidarity. In the absence of such networks, the largely negative coping strategies farmers adopted with regard to taxes and prices could hardly offset the detrimental influences on their revenue, which originated above all in the skewed organization of the market. Although the coping strategies described above could potentially provide these farmers with an important endowment of social capital, our analysis suggested that they were confronted with an 'excess of community' – that is, a situation of trust that extended only to members of the same small group.[23] This situation was in sharp contrast to the vertical ties that enabled commercial intermediaries and other local elites to develop favourable economic relationships *outside* their small circles of relatives and friends.

This analysis suggests that a differentiation should be made between various forms of social 'capital', i.e. between social ties that impede and those that enable households to develop favourable relationships to confront their problem of food insecurity. While farmer households commonly are characterized by horizontal social ties (which impede them from confronting the negative consequences of conflict), commercial intermediaries are often able to generate opportunities from the war by forging autonomous social links with outside contacts and markets.

Land: Moving the problem

Another structural factor that influences food systems, particularly during protracted crises, is access to land. As indicated previously, the mediating structures that regulate this access are a critical factor for inhibiting or facilitating livelihood capabilities and choices. There is, for example, an important difference between private freehold titles and customary land rights, which result in different structures conditioning livelihood strategies. This difference becomes especially evident during protracted conflict crises.[24]

In Beni-Lubero, the access of poor households to land began to be particularly problematic beginning with the introduction of the Bakajika land law in the 1960s and 1970s. The privatization of land ownership during this period led to a growing accumulation of land titles in the hands of rural 'capitalists', who often worked in compliance with local ruling elites in the Mobutu administration. The Congolese civil war aggravated this situation by forcing peasants to cling to their existing land titles. Due to continuing population pressure large families had to cope with ever-shrinking plots of land, while capitalist traders and politicians continued to buy up large tracts of land as investments. This pushed many farmers to migrate from their original homesteads towards other regions, including the region's national parks.

The national park

One common strategy for households confronted with land problems consisted of the invasion of the region's national park. According to local estimates, more than 400 ha of Virunga National Park (situated in Rutshuru and Lubero) were gradually invaded by local villagers from both Beni and Lubero (Butonto, 2004). Because of its favourable location in Africa's Rift Valley, the park offered perfect conditions for the production of subsistence and commercial crops, especially since the park's wildlife population was chased out long ago by roaming rebels and militias. An increasing number of households confronted with the depletion of their traditional assets used this alternative as an escape from destitution. Great quantities of rice, maize, soya, bananas and manioc were cultivated in the northern and western areas of Virunga National Park by households from the region around Butembo (Bunyuka, Bulambo and Ishale), and from central and southern Lubero. The local agricultural output led to an alternative commercial circuit consisting of trade in agricultural goods through and from the park (see Box 9.1). The lack of an efficient environmental protection mechanism combined with the roaming militias facilitated use of park land as an economic alternative for North Kivu's population, which would have been doomed otherwise by the erosion of their land rights. However, the alternative was blocked increasingly by the same actors and organizations responsible for the destruction of local assets (see Box 9.1).

Western Lubero

Another strategy of local households – especially from central Lubero – was to move west towards the forest. For at least two decades the population of central Lubero had been confronted with growing pressure on its existing land titles. During the mid-1990s, a series of interventions by both local NGOs and the Catholic Church encouraged these central Lubero households to resettle to several 'virgin' sites – including Katimbya, Muhangi and Musasa – in the western Lubero forest area. This relocation was explained as having a number of significant advantages both in terms of land access and agricultural productivity. In Vuyinga (in western Lubero), households obtained on average a 10.3 ha arable field for a local inhabitant and a 5 ha field for an 'immigrant'; in Masereka (in central Lubero), the size of plots was generally 0.3–1 ha (Vahamwiti Mukesyayira, 2005). While more than half of the land titles in Masereka were considered 'contested', in Vuyinga it was only 12 per cent. While 80 per cent of households in Vuyinga were considered to have secure land rights, only one-third could claim the same in Masereka (Vahamwiti Mukesyayira, 2005). In addition, the population density in the western forest area was substantially lower (69.4 inhabitants per km^2) than in central Lubero (253 per km^2). However, by 2015, the pressure on arable land in the forest area will have doubled (Vahamwiti Mukesyayira, 2005). If conditions remain

Box 9.1 Fishing in troubled waters: The exploitation of Lake Edward

Lake Edward, in eastern Lubero, has great potential, but its fishing reserves have been greatly overexploited. Its overall output fell from up to 12,000 metric tons (mt) per year in 1954 to 3,000 mt in 1989 (Vakily, 1989); the daily production per *pirogue* (canoe) dropped from between 1,500 and 3,000 fish in 1984 to 1996 to just 30 fish in 2006 (VECO RDC, 2006). The reasons for this radical decline may be found in the institutional disorganization surrounding the lake's exploitation: following independence from colonialism, an amalgam of actors and organizations emerged to compete for access to the lake's reserves. One of these organizations, COPEVI (*Coopérative des Pecheurs de Virunga*), disintegrated into several fisheries as a result of intra-organizational and military competition. (Because of the war, Lake Edward's two largest fisheries, Vitshumbi and Kyavinyonge, each fell under the control of a different rebel militia.) Another organization responsible for the situation was the ICCN, the Congolese institute for the protection of national parks: although its role was to regulate access to the protected part of Lake Edward, its agents controlled a number of private fisheries, where individual fishermen 'monitored' access on the agents' behalf (Butunto, 2004).

Confronted with this organizational disarray, the population bordering Lake Edward adopted a series of alternative coping strategies to continue providing for their livelihoods. One of these was for farmers, customary heads and local NGOs to form discussion platforms to seek access to the unexploited reserves of Virunga National Park in order to cultivate the land. A direct consequence of the new agricultural output from the park was an explosion of petty trade in the area. While rice had traditionally been imported to the area, former fishermen began sending rice they grew in the park to the market in Goma. Soya and coffee – and sometimes ivory and bushmeat – were exported illicitly to Uganda, using the same *pirogues* formerly used for fishing. Another distribution chain was that of salted fish and *djoro djoro* (fish waste) from Uganda: large quantities were imported on a weekly basis to places like Goma and Butembo – again using fishing boats as a means of transport. The boom in the local transport business stimulated the production of *pirogues* in the fishing villages, which became the centre of transborder trade.

Increasingly, however, the growing adaptiveness of Lake Edward's population became a thorn in the eye of 'environmental' agents, who saw their position threatened by such independent responses. The local discussion platforms in Virunga National Park became the target of a slander campaign that portrayed them as a threat to the DRC's national reserves. At the same time, the park attracted a series of free-riders – among others the customary landlords, who claimed the right to demand tributes from immigrant farmers. The potential resilience of Lake Edward's population was blocked once again by the same local institutions complicit in the depletion of scarce local resources.

Since 1990 only three projects attempted to help the fishing villages on the edge of Lake Edward. The first project, to reinforce the production capacities of COPEVI, was thwarted by the various rebellions erupting in Masisi and Lubero in 1993; the two freezers and a freezer lorry that were delivered subsequently disappeared. For the second project, in 2004–2005, FAO distributed fishing nets to the two main fisheries (Kyavinyonge and Vitshumbi) with the aim of assisting vulnerable women. With practically no fish left in the lake, the nets were quickly sold to buy two outboard motors to be used for transborder transport and the project was abandoned due to insufficient local support. The third project, still in progress in early 2007, involved constructing a pipe to supply the village of Kyavinyonge with drinking water, financed by VECO-Congo.

(written with Nzangi Muhindo Butonto)

unchanged, by 2020 years the population pressure on land in western Lubero will have become very similar to that in the immigrants' region of origin, with many of the same problems.[25] Because of this, many (especially old male) immigrants have chosen to return to their home areas, or at least leave their families while they cultivate in this inhospitable forest area.

Several elements support the hypothesis that pressure on land will soon be as strong in western Lubero as in central Lubero (i.e. the problematic area of Masereka). First, there is an almost total lack of infrastructure in the forest of western Lubero: there are no roads, and almost no schools or health centres. Immigrant peasants have to spend much of their energy toiling on roads and making field tracks accessible. Lack of adequate infrastructure leads to almost constant labour fatigue, which negatively influences health. Another problem is shelter. Constructing provisional homes involves heavy labour and the lack of decent shelter makes it very difficult to store agricultural products. The ongoing need to repair shelter involves a considerable waste of time that would otherwise be spent on agricultural work. To spend as little time as possible on the construction of their houses, farmers usually build small huts with palm leaves or ferns, using reeds or leaves for doors instead of wood (which can be found only very far away), but every two years the entire home must be rebuilt. A number of environmental factors also work to the disadvantage of immigrant farmers, who cite as most significant soil infertility and plant diseases. The soil is argillaceous in most of western Lubero, and manioc remains virtually the only crop that can withstand local environmental conditions. Without adequate farming techniques, monoculture leads to leached soil. The results are a rise in destructive plant diseases (for example mosaïc), and a drop in agricultural production.

In addition to these problems with infrastructure and environment, 'institutional' factors again enter the picture. In several cases, immigrant farmers referred to the obstructionist attitudes of local customary chiefs, who fiercely resisted the 'intrusion' of these new inhabitants into their fiefs. As in their home region, immigrant farmers had to pay tributes to obtain access to a field from these customary landlords. As a way of controlling the land tenure of these new occupants (and also to protect their own extended kin), landlords implemented a series of methods that seriously impeded immigrants from developing a stable economic relationship with their host region. One method was to sell off any fields that lay fallow: because landlords considered such fields to be 'unoccupied', they automatically concluded that they could sell them to others. As a result, immigrant farmers were increasingly forced to look for land in remote areas, where they cultivated ever-smaller fields. In certain cases, landlords even destroyed fields before farmers had a chance to harvest their crops.

Other aspects of 'culture' also intervened. Some customary landlords prohibited farmers from planting certain crops because of supposed 'traditional' directives. On more careful examination, such prohibitions seemed to be mainly economically motivated. By invoking 'culture', landlords prevented immigrant farmers from challenging their political and economic positions, by, for example, planting commercial crops that the peasants could have sold on local markets. The result of such 'institutional' pressure was that instead of benefiting from more secure land rights in the new area, immigrant farmers ended up cultivating the same size plots as in their home regions in central

Lubero. Finally, this equal pressure on land titles, combined with a lack of infrastructure and favourable environment, meant that immigration to the area remained largely temporary: the immigrants that moved to western Lubero to cultivate plots were mostly old men who left their entire families behind in central Lubero to wait for better times to come.

Food security interventions

In spite of the obviously structural dimension of Beni-Lubero's food crisis, by early 2007 aid interventions for the area were still being guided by a rather narrow interpretation of food security. On the one hand, there had been a withering of the emergency response given the 'neither war nor peace' situation that characterized Beni-Lubero from late 2004. The relative calm that settled over most parts of the two territories meant that most humanitarian programmes either closed down or reduced their activities to a minimal level. This policy shift was apparently inspired by fatigue on the part of international donors, who were pulled increasingly towards more 'serious' crises such as that in Ituri.

On the other hand, most agencies continued to be reluctant to implement more development-oriented programmes that would seek to stabilize livelihoods affected by the war.[26] This reluctance was reinforced by continued insecurity, particularly in western and southern Lubero, where several militias continued to operate. Furthermore, the actions of MONUC and the FARDC *against* these irregular militias were increasingly blocking the efforts of development agencies. In southern Lubero, for example, the operations of MONUC and FARDC against the Rwandan Hutu militias led to a dispersion of those forces to the north; the result was an increase in retaliations by the militias against civilians. The presence of FARDC forces, rather than guaranteeing security, contributed to greater insecurity in these areas. This was especially the case in the northern and central parts of Lubero, which were increasingly inaccessible to international agencies and civilians alike.[27]

The lack of a long-term perspective was also reflected in the dearth of information regarding livelihood strategies and opportunities: very little data were being gathered on the medium- and long-term features of Beni-Lubero's food crisis. Most information collection was being done in relation to immediate needs assessments and the presence of IDPs.[28]

This absence of a long-term perspective was aggravated further by a general lack of coordination between international agencies as well as a lack of opportunities for local agencies to participate in externally-financed programmes. Although FAO had initiated food security meetings that gathered together several agencies on a regular basis, the meetings were used to discuss mainly immediate matters of food and asset distribution rather than concerted action or coordination of activities. Little or no exchange of information existed about the execution of overlapping programmes. Despite a shift towards more community-based partnerships on the part of some intervening organizations, the role of local organizations remained limited mostly to the execution of international programmes. There was only one organization in the entire region that focused on the problem of land access: with the help of VECO, SYDIP (Syndicat des Intérêts des Paysans) had developed a programme of conflict resolution directed towards customary landlords. Table 9.3 provides an overview of food security interventions in Beni-Lubero during 2005, with a discussion of advantages and disadvantages with reference to FAO's twin-track approach.[29]

Conclusions

This chapter discussed the impact of protracted conflict on the food security situation in Beni-Lubero, a region demonstrating elements of both war and peace. While since December 2004 there had been relative calm due to the application of several cease-fire agreements, it still had a number of significant pockets of insecurity. Among these were southern Lubero and the western Lubero forest, both characterized by general insecurity and lack of basic infrastructure.

The 'neither war nor peace' situation that had settled over the region posed several challenges to both policy analysis and response. Post-conflict situations such as the one in Beni-Lubero are very often characterized by a *policy gap* between the constantly changing nature and causes of food insecurity on the one hand, and policy response on the other (see Pingali et al, 2005). Although it is increasingly acknowledged that food insecurity in such cases is caused by a complex and dynamic set of causes, responses to food crises still tend to be driven by a one-dimensional understanding of the phenomenon. One major element that is often disregarded is the 'institutional' context, the 'mediating structures' that either block or facilitate livelihoods' effective adaptations to such crises. These changes were illustrated through a discussion of two main problems affecting Beni-Lubero's households: access to the market and access to land (or fishing reserves, in the case of Lake Edward).

The analysis of household responses to the Beni-Lubero food crisis suggested that household coping strategies had not led to greater resilience for rural livelihoods. On the contrary, the study of livelihood diversification indicated a kind of 'back to the future' scenario, in which the structural causes of poverty and vulnerability had been largely reconfirmed. While the war had worked to the advantage of a small but powerful rural capitalist class, attempts by poor farmer households to diversify their livelihoods were blocked by various mediating structures. The lack of effective adaptation mechanisms could be explained in part by the lack of linkage between households and networks that reached beyond their communities. The 'excess' of community existing at a horizontal level was reinforced by various 'gatekeepers' (intermediaries in the case of the market and obstructive landlords/state agents in the case of immigrant farmers) that continued to block the access of most households to opportunities and assets. A cluster of institutional factors further limited

Table 9.3 Overview of food security interventions in Beni-Lubero during 2005 using FAO's twin-track approach

Responses	*Partners*	*Key objectives*	*Advantages*	*Disadvantages*
Food aid distribution	International agencies, often through local associations (Oxfam, Solidarité, Caritas, World Vision, CESVI, Première Urgence, Médecins Sans Frontières, NRC, Save the Children, SODERU, etc.)	Address immediate food needs	Focus on direct, immediate access to food through increased availability (food aid, enhancing food supply to the most vulnerable) and through increased food access (nutrition intervention programmes); strengthening of the labour market	Continuing dependency of the poorest; vulnerability to military aggression (south Lubero); failure to address underlying access problem
Distribution of seeds and tools	International agencies and local associations (Ibid)	Diversify crops; facilitate access to seeds and tools; increase food production	Focus on direct, immediate access to food through increased food availability (seed input relief) and on rural development through stability efforts (improvement of rural food production)	Problem of association membership (only members gain access to seeds); failure to address land access problem (central Lubero); disruption of distribution mechanisms based on microcredit.
Provision of technical support	International agencies, often through local associations (AAA, FAO, ASSOPELKA, APETAMACO, etc.)	Improve production of food; strengthen capacities to deal with diseases	Focus on rural development/ productivity enhancement through increased food availability (improvement of rural food production and investment in rural infrastructure) and food stability (diversification of agriculture and employment, monitoring of food security and vulnerability, and development of risk analysis and management)	Insufficiency; failure to address land problem (west Lubero)
Collective labour (agricultural labour rotating mechanisms)	International organizations in collaboration with local associations or 'village committees' (AAA, FAO)	Facilitate access to land for landless farmers; increase food production	Focus on rural development/ productivity enhancement through increased food availability (improvement of rural food production) and access (to land)	Lack of grassroots participation (similar to cooperatives); often no production increase (southern Lubero)

Responses	*Partners*	*Key objectives*	*Advantages*	*Disadvantages*
Rehabilitation of rural infrastructure (roads, etc.)	International agencies (AAA)	Rehabilitate roads; facilitate access to local markets; increase food and cash distribution (food for work/ cash for work)	Focus on direct, immediate access to food through increased access (food/ cash-based transfers, and asset redistribution if cash based) and focus on rural development through increased availability (investing in rural infrastructure)	Dependency on food aid (in the case of food for work)
Conflict resolution	Local associations (SYDIP)	Mediate in land disputes; strengthen the legal position of farmers	Focus on rural development/ productivity enhancement through increased access (to land) and food stability (dealing with structural causes of food insecurity)	Not all customary land rights necessarily advantageous to the poor
Introduction of micro-credit systems	Local associations (SYDIP, LIDE (Ligue de Developpement), UWALU, etc.)	Facilitate the revival of rural financial systems	Focus on rural development/ productivity enhancement through increased access (reviving financial systems) and through stability (reviving of access to credit systems and saving mechanisms)	Problem of membership (only members gain access to credit); vulnerability to fraud
Commercial cooperatives	International organizations, in collaboration with local associations (VECO, COOCENKI)	Facilitate households' access to the market	Focus on food availability by enabling market revival, and on access (enhancing access to assets)	Lack of grassroots participation; experience of elite appropriation and fraud (entire region).

Source: Author compilation

peasants' agricultural production and revenue, leading to continued poverty and vulnerability. These factors could be characterized as a lack of institutional *organization*, leading to a depletion of household assets, and a lack of ensured *access*, expressed in the monopolistic actions of several 'gatekeepers' (customary landlords, environmental agents or commercial intermediaries) that blocked opportunities for a majority of the households.

The lack of ensured access of smallholders to arable fields was still a key determinant of Lubero's food crisis. To confront the growing land problem, a number of organizations intervened in favour of the relocation of households affected by land conflicts in central Lubero to more sparsely populated areas to the west. While at first glance these interventions appeared to offer some significant advantages in terms of availability of food and access to food, this chapter concluded that such relocation offered only a temporary solution. The scarcity of infrastructure and the obstructive attitude of local landlords (who were generally hostile toward the 'intrusion' of newcomers into their area) meant that the peasants' efforts to enhance their livelihoods were largely in vain. If conditions remained unchanged, within a decade or two the outcomes of Beni-Lubero's food crisis were likely to look very similar in both central and western Lubero.

The chapter comes to a similar conclusion regarding the exploitation of Lake Edward. Once the fishing reserve of the entire province of North Kivu, the output of Lake Edward had declined radically since at least the 1980s. The reasons for this included institutional disorganization and lack of an efficient protection mechanism for the lake's fishing potential. In the absence of such a framework, poor peasant households were forced to seek alternatives, including the invasion of the region's national park. However, households were confronted with the same problems in the park land as they encountered on the lake: obstruction by the agencies that had contributed originally to the decline in the households' food security. In the case of the park, it was the environmental services, complicit in the destruction of Lake Edward's productive output, that prevented households from making a living in the forest.

The chapter also examined access to markets, and concluded that farmers confronted with acute crisis situations do not usually withdraw completely into subsistence. On the contrary, the revenue acquired from the sale of agricultural products constituted a central element of local food security. However, peasant households faced two major problems regarding agricultural trade. The first was a lack of *accessibility*: due to lack of local infrastructure such as roads and transport facilities, farmers encountered major difficulties in getting their products to market. The second problem was that of *access*: organizational factors had significant negative impact on peasants' agricultural revenue. One factor was the various intermediaries who monopolized the access of agricultural households to the market and imposed the prices to be paid to them. Another factor was 'taxes', the various tributes and harassment peasants had to face on their way to market.

Despite the preponderance of these 'structural' factors, most intervening agencies preferred to focus on inputs related to the immediate availability of food, such as food aid and free seeds, or the rehabilitation of agricultural infrastructure. This is of course understandable if one takes notice of the history of violence and humanitarian emergency that characterizes the eastern Congolese region in general. One potentially positive outcome of this focus on short-term relief also consists of a modest agricultural recovery, which results in part from improved access to seeds and agricultural equipment. However, this outcome is unlikely to have positive effects in the longer term if it is not coupled with a more adequate response to the institutional dimension of Beni-Lubero's food crisis once the peak of conflict-induced insecurity lies behind us.

Some direct recommendations emerged regarding both the analysis and resolution of food security crises, particularly in situations where a combination of protracted conflict and collapse of state institutions hampers the access of poor households to vital livelihood assets. Such situations do not necessarily entail the collapse of society in general: households continue to offer original responses to protracted political and economic crises, and their survival strategies will very likely influence the choices of ruling elites in addressing their respective policy agendas. What determines the transformation of household *coping* into resilient *adaptation* to external shocks, however, remains the intermediary and regulating institutions that are often controlled by these same (formal or informal) ruling elites. Often the problem is not a scarcity of resources as such but rather the management of those resources by a series of exploitative institutions and networks, whose very existence often depends on limiting the access of poor households by way of intermediary checkpoints or gatekeepers.

The way to ensure better access of poor households to food and economic assets is to ensure their increased participation in food security interventions, which can be achieved by gradually shifting the focus of interventions from short-term food *availability* to longer-term issues of *access* and *accessibility*. This shift will have limited impact, however, until serious attempts are made to address the root causes of Beni-Lubero's ongoing food crisis: unequal access to both land and markets.

References

Ballentine, K. and Nitzschke, H. (eds) (2005) *Profiting from Peace: Managing the Resource Dimensions of Civil War*, Lynne Rienner Publishers, Boulder, CO.

Butonto, M. N. (2004) *L'exploitation du Lac Edouard et son impact sur la situation économique des armateurs*, Faculté des Sciences Economiques, Université Catholique du Graben, Butembo (DRC).

Collier, P. (2003) *Breaking the Conflict Trap: The Civil War and Development Policy*, World Bank, Washington DC and Oxford University Press, Oxford.

Collinson, S. (ed.) (2003) *Power, Livelihoods and Conflict: Case Studies in Political Economy Analysis for Humanitarian Action*, HPG Report 13, Overseas Development Institute, London.

Davies, S. (1996) *Adaptable Livelihoods: Coping with Food Insecurity in the Malian Sahel*, Palgrave MacMillan, London.

Davies, S. and Hossain, N. (1997) *Livelihood Adaptation, Public Action and Civil Society: A Review of the Literature*, Working Paper 57, Institute of Development Studies, University of Sussex, Brighton.

Ellis, F. (2000) *Rural Livelihoods and Diversity in Developing Countries*, Oxford University Press, Oxford.

Fairhead, J. (1992) 'Paths of authority: Roads, the state and the market in Eastern Zaire', *European Journal of Development* 4 (2): 17–35.

Flores, M., Khwaja, Y. and White, P. (2005) 'Food security in protracted crises: Building more effective policy frameworks', *Disasters*, 29 (S1): 25–51.

Granovetter, M. (1973) 'The strength of weak ties', *American Journal of Sociology*, 78 (6): 1360–1380.

Hyden, G. (1980) *Beyond Ujamaa in Tanzania: Underdevelopment and an Uncaptured Peasantry*, University of California Press, Berkeley, CA.

Jackson, S. (2003) 'Fortunes of war: the coltan trade in the Kivus'. In S. Collinson (ed.) *Power, Livelihoods and Conflict: Case Studies in Political Economy Analysis for Humanitarian Action*, HPG Report 13, ODI, London.

Laurent, P. J., Mafikiri Tsongo, A. and Mathieu, P. (1996) *Mouvements de Populations, Cohabitations Ethniques, Transformations Agraires et Foncières dans le Kivu Montagneux*, Institut d'etudes du développement, Louvain-la-Neuve.

Levine, S. and Chastre, C. (2004) *Missing the Point. An Analysis of Food Security Interventions in the Great Lakes*, HPN Network Paper No. 47, Overseas Development Institute, London.

Longley, C. and Maxwell, D. (2003) *Livelihoods, Chronic Conflict and Humanitarian Response: A Synthesis of Current Practice*, ODI Working Paper No. 182, Overseas Development Institute , London.

MacGaffey, J. (1987) *Entrepreneurs and Parasites. The Struggle for Indigenous Capitalism in Zaire*, Cambridge University Press, Cambridge.

Mafikiri Tsongo, A. (1994) *La problématique foncière au Kivu montagneux (Zaïre)*, Cahiers du CIDEP, No. 21 (September), L'Harmattan, Paris.

Maxwell, D., Watkins, B., Wheeler, R. and Collins, G. (2003) 'The Coping Strategies Index: A tool for rapidly measuring food security and the impact of food aid programmes in emergencies', paper presented at the FAO International Workshop on Food Security in Complex Emergencies: Building Policy Frameworks to Address Longer-Term Programming Challenges, 23–25 September 2003, Tivoli, Italy.

Newbury, C. (1984) 'Ebutumwa bw'emihogo. The tyranny of cassava: A women's tax revolt in eastern Zaire', *Canadian Journal of African Studies* 18 (1): 35–54.

OCHA (Office for the Coordination of Humanitarian Affairs) (2005) *Evaluation d'urgence de sécurité alimentaire et de securité des strategies de vie des ménages au Nord Kivu, Goma*, (April), OCHA, Geneva.

Pingali, P., Alinovi, L. and Sutton, J. (2005) 'Food security in complex emergencies: Enhancing food system resilience', *Disasters* 29 (S1): S5–S24.

Pottier, J. and Fairhead, J. (1991) 'Post-famine recovery in highland Bwisha, Zaire: 1984 in its context', *Africa* 61 (4): 437–470.

Remotti, F. (1993) 'Società, matrimoni, potere' (Volume I). In F. Remotti (ed.) *Etnografia Nande* (2 volumes), Il Segnalibro, Turin, Italy.

Richards, P. (ed.) (2005) *No Peace, No War. An Anthropology of Contemporary Armed Conflicts*, James Currey, Oxford.

Vahamwiti Mukesyayira, J. C. (2005) *Foncier, population et pauvreté en chefferie des baswagha au Nord Kivu*, Faculté des Sciences Economiques, Université Catholique du Graben, Butembo, DRC.

Vakily, J. M. (1989) *Etude du potentiel halieutique du lac Idi Amin*, European Economic Commission, Brussels.

VECO RDC/Réseau WIMA (2006) La *sécurité alimentaire à Kyavinyonge (lac Edouard)*, Etude MARP, mimeo.

Vlassenroot, K. (2002) The Making of a New Order. Dynamics of Conflict and Dialectics of War in South Kivu (DR Congo), unpublished PhD thesis, University of Ghent, Belgium.

Vlassenroot, K. and Raeymaekers, T. (2004a) *Conflict and Social Transformation in Eastern DR Congo*, Academia Press, Ghent, Belgium.

Vlassenroot, K. and Raeymaekers, T. (2004b) The politics of rebellion and intervention in Ituri: The emergence of a new political complex?, *African Affairs* 103 (412): 385–412.

Vlassenroot, K., Ntububa, S. and Raeymaekers, T. (2006) *Food Security Responses to the Protracted Crisis Context of the Democratic Republic of the Congo*, report for FAO Food Security Information for Action, FAO, Rome, ftp://ftp.fao.org/docrep/fao/009/ag307e/ag307e00.pdf [accessed 16 January 2007].

Vwakyanakazi, M. (1987) 'Import and export in the second economy in North Kivu'. In MacGaffey, J. (ed.) *The Real Economy of Zaire. The Contribution of Smuggling and Other Unofficial Activities to National Wealth*, University of Pennsylvania Press, London.

Woolcock, M. (1998) 'Social capital and economic development: Toward a theoretical synthesis and policy framework', *Theory and Society*, 27 (2): 151–208.

WPF (World Food Programme) (2004) *Rapport sur l'analyse et la cartographie de la vulnérabilité à l'insécurité alimentaire au Nord Kivu, Goma*, (May), WFP, Rome.

CHAPTER 10

Land tenure, conflict and household strategies in the eastern Democratic Republic of the Congo

Koen Vlassenroot

Abstract

The problematic relationship between land tenure, food security and conflict has generated a considerable body of research. Land disputes are increasingly recognized as dynamic processes generated by, sometimes perceived, land tenure insecurity. Conflicts, however, can also lead to intensified struggle for land, especially when political and military elites seek to consolidate their power base and reward their supporters by extending control over land as part of their war strategies. This chapter looks at the links between local mechanisms of land distribution and conflict in the Eastern Democratic Republic of Congo. It focuses on a number of regions where local patterns of land use and land access can be identified as key dynamics behind local tensions and disputes.

Introduction

The intricate relationship between land tenure, food security and conflict has recently generated a considerable body of research. It is more and more often recognized that 'land access constitutes one of the more problematic and volatile facets of societal relations during and subsequent to armed conflict' (Unruh, 2003). Land disputes are also increasingly understood as dynamic processes that are generated by (perceived) land tenure insecurity (Maxwell and Wiebe, 1999; Deininger, 2003; Broegaard, 2005; Huggins and Clover, 2005). This tenure insecurity is caused in many cases by shifts in the local rights and institutions that govern access to and use of land. When shifts in the institutional context and governance of land tenure result in insecure and limited access to land for large sections of society, they become structural causes of poverty, food insecurity and conflict. Conflicts, however, can also lead to intensified struggle for land, especially when politico-military elites seek to consolidate their power base and reward their supporters by extending control over land as part of their war strategies. In these cases, land is transformed from a structural source of poverty and crisis into a 'resource of conflict'.

The recent history of the DRC offers a good illustration of the links between shifts in the structural context of land distribution and violent conflict. The Congolese war has often been presented as a prominent example of how violent struggles over natural resources have shaped internal warfare. While some have noted a 'commoditization of war' (Jackson, 2003), others have pointed to how natural resources have shaped the power strategies pursued by the different belligerents (United Nations Security Council, 2001). Recent analysis, however, has revealed that in eastern DRC local-level disputes over land have tended to dominate socio-economic competition, pitting entire communities against one another and leading to regular outbursts of violence (Pottier, 2003; Vlassenroot and Huggins, 2005). Since the start of the Congolese war, these disputes have become linked increasingly to the larger, regional struggle for economic control and politico-military power, turning land into an important asset for both non-state armed groups and national armies involved in the DRC war. The conditions offered by the war context have motivated a class of local and regional businessmen, politicians, traditional authorities and landowners to develop new strategies to increase their control over tracts of land. At a societal level the competition for land between local and regional actors has also stirred up violent clashes between families, clans and entire ethnic communities and has affected existing mechanisms of land distribution.

This chapter provides a systematic description of the particular links between local mechanisms of land distribution and conflict in the territories of Walungu (South Kivu), Masisi and Lubero (North Kivu). In each of these regions, local patterns of land use and land access were identified as key dynamics of tension and dispute. The territory of Walungu (South Kivu) has witnessed increased competition for land because of growing demographic pressure and high levels of insecurity caused by the Congolese war. The territory of Masisi (North Kivu) has been the scene of local struggles for land since the 1960s. In 1993, increased pressure on land generated ethnically motivated violence and militia activities, a conflict that by early 2007 had still not been resolved. The central highlands of Lubero (North Kivu) are faced with increasing population density and the reduction of market opportunities (partly as a result of the war). These conditions have provoked growing competition for land. In the regions under study, we observed that a context of war and struggle for land had forced households to develop alternative strategies to secure access to food. We argue that these food security strategies at the household level eventually induced important transformations of local food systems,[1] which included shifts in consumption patterns and modes of food production. We analyse the consequences of these shifts and how food security interventions have dealt with these structural causes of food insecurity and transformations in local food systems.

This analysis is based on extensive field work in eastern DRC carried out in November and December 2005. The field work concentrated on the larger political economy (or institutional context) that conditions people's access to land, as well as on people's claims and strategies to obtain access to land to use it. It focused on the different mechanisms that regulate households' entitlements to food and on the impact of household coping mechanisms on local food systems. Empirical data were gathered through participative research methods and interviews with key stakeholders. The assessment of household strategies and responses included an evaluation of both coping strategies (Maxwell et al, 2003) and longer-term adaptation strategies.

The first part of the chapter focuses on the institutional context regulating local access to land in the various case studies;[2] it examines how shifts in this context have intensified competition for land access. The second part analyses the impact of land scarcity on food security and rural households. The third part evaluates household strategies in a context of war and food insecurity. The final part explores the impact of food security interventions.

The dynamics of land tenure in DRC

As in many other parts of Africa, land tenure systems in eastern DRC have undergone a number of changes during recent history. Colonial land policies and the use of land as a political asset to reward members of the ruling political class after independence led to structural change in rural society. In the most densely populated areas they increasingly limited people's access to land.

Before the colonial conquest, large parts of eastern Congo were characterized by stratified social structures, which organized use of the available space by distributing access rights to land held under customary ownership in return for rent. These contracts could be described as 'institutions': they not only regulated access to land but also legitimized an entire social organization by integrating all persons living within a given region into a local network of dependent relations (Van Acker, 2005), thus granting them an identity within local society. Small tributary states and chiefdoms headed by a *mwami* (customary chief) and circumscribed by the boundaries of clans or ethnic communities formed the institutional framework for this highly stratified and patriarchal social structure. In this structure, property rights were limited to use rights in return for the payment of a tribute to the representative of the *Mwami*. According to this principle, access to land depended on an initial payment of tribute to the chief. Once this tribute was paid, the peasant obtained user rights over a part of the customary land, though in practice the peasant regularly had to renew his tribute. The main objective of these tributary systems was to recycle the rents paid as tribute by those given access to land and to enable the nobility to extract the surplus generated by the labour of farm households. The different rents paid by the producers and redistributed to higher levels of the hierarchy had to guarantee the sustainability of the network of dependent relations and thus, of the existing social order. The result of this system was a complex structure of rights where nobody had complete property rights, but few – if any – had no rights at all: at the top the custodian of the tribal land

Box 10.1 The customary tribute system

In Congo's traditional society, tributes were the economic backbone of the customary order and were 'redistributed upwards in an elaborate system of dependency that configures the collective management of economic uncertainty' (Van Acker, 2005). The main objective of these tributary systems was to recycle the rents paid as tribute by those given access to land, and to enable the nobility to extract the surplus generated by the labour of farm households. In addition, this tribute system gave meaning to the local social order, as it was the basis of a very rigid social pyramid. The set of institutions that structured this social organization offered farmers access to land, the necessary means of social integration and customary protection but at the same time tied them to their villages, kept them firmly subjected to their chiefs and consolidated their very dependent position (Sosne, 1979). The rents paid by the producers and redistributed to higher levels of the hierarchy guaranteed the sustainability of the network of dependent relations and thus of the existing social order. The result of this system was a complex structure of rights where nobody had complete property rights, but few – if any – had no rights at all: at the top was the custodian of the tribal land (mwami) and at the bottom peasants that paid tribute. For a peasant family, the system traded social integration (and hence security) for loyalty and tribute to the mwami, who received power in exchange for granting non-alienable use rights over the customary domain.

(*Mwami*) and at the bottom peasants that paid tribute without receiving any rights.

The socially integrative aspect of these customary land systems came under pressure for the first time during colonialism. After establishing territorial limits to the traditional rural order, the Belgian colonial powers initiated a dual system of property rights. The colonial administration declared all vacant land to be property of the colonial state and introduced a system of land registration and private ownership. This allowed it to regulate the access of the colonialist commercial class to these vacant lands so they could be turned into plantations; other parts of the territory were turned into wildlife parks and anti-erosion forests. The legitimacy of the existing customary land tenure system was recognized only for land already under the practical control of the traditional authorities, thus limiting any further expansion of customary lands. In Walungu, by 1956 already 9.7 per cent of the land had come under concession to European planters (DeBacker, 1958). Agricultural production at these plantations was based on forced labour, which did not offer a very viable alternative to the traditional peasant. As average wages were extremely low, this colonial policy provoked the start of a process of peasantization and proletariatization, and strongly reduced access to land for peasant families living in densely populated regions such as Masisi and Walungu. Hecq and Lefèbvre (1959) estimated that in order to have sufficient nutrients a four-member family needed the produce from a plot of 1.2 ha and a supplement of palm oil and salt. By 1959, households in Kabare (South Kivu) occupied on average less than 1 ha.

After independence, the mechanisms of land access were further complicated and confused by the 1973 General Property Law, which declared all land (including land under customary control) to be property of the state. This new legal framework had three major effects. First, it undermined the traditional system of reciprocal patron–client relations embedded in a customary framework, to the advantage of a new type of interaction based on state patronage. Secondly, the inclusive traditional social order, even if leading to social stratification, was replaced by a process in which proximity to the political centre became a premium condition for the accumulation of wealth. Thirdly, property structures drastically changed because the new land law became part of a policy of opportunistic nationalization of plantations. Beginning in the 1970s, political loyalty was rewarded with the distribution of land, which became a crucial asset within the new patrimonial system. Members of new, opportunistic alliances benefited from the redistribution of nationalized plantations and customary land. As a result, the majority of land titles came under the control of a small number of landowners, while most small farmers were pushed into a position of insecure land titles (Tsongo, 1994; Mararo, 1997).

The extraction of land from the system of customary ownership also had an effect on the traditional social structure. Capitalizing the rents embedded in land ownership eroded the web of mutual dependency that was built on the careful extraction and (re-)distribution of these rents; it also led to a different qualitative use of the available space. Customary chiefs came to be integrated into the administrative structures of the state. As a consequence, their position was no longer based on their customary powers of land distribution but depended on their integration into state-based networks of patronage. In order to consolidate their new position, the chiefs weakened the customary land use rights of farmers and facilitated the purchase of titles to customary land. Fieldwork in Walungu revealed that traditional authorities then developed a number of strategies that limited the access of local households to land. One strategy was to dispute the legal status of farmers' land use rights. Another was to question whether tributes had been paid regularly; a failure of payment could be interpreted as customary treason, allowing customary chiefs to seize the land (Van Acker, 2000). In addition, statements of vacancy and registration were delivered without informing the local population, and land that traditionally fell under the direct control of customary chiefs (such as marshlands) was privatized and sold to rural entrepreneurs or churches. The effects of this practice can hardly be underestimated, as these wetlands were traditionally considered to be property of the entire community and accessible to farmers only in cases of economic or environmental crisis.

In Masisi, the immigration of large numbers of Banyarwanda from neighbouring Rwanda further complicated local patterns of land distribution. Following the introduction of the new property law in the 1970s, one-third of the available space came under the control of a small class of landowners and was used mainly for commercial ranching and plantations of coffee, tea or pyrethrum (a type of chrysanthemum cultivated primarily for use as pesticide). Clientelistic relationships between traditional authorities, politicians and rural capitalists not only limited the land claims of local peasants, but also provoked struggles between ethnic communities as large tracts of land were extracted from the local customary system to the advantage of a newly settled 'allochthonous' elite. Because of their easy access to the inner circles of Mobutu's patronage network, a small group of Banyarwanda (mainly Rwandan Tutsi refugees that had arrived between 1959 and 1963), as well as a number of political allies of Mobutu, gained control over the land concessions and plantations, which had belonged to white settlers until the 1973 'Zairianization'. This control was often accomplished through controversial transactions and with the direct involvement of indigenous customary chiefs and the Services Fonciers Provinciales (provincial services responsible for land issues). The new patterns of land distribution drastically transformed the existing property structure and consolidated the differences between new farming systems oriented towards 'the market' and traditional subsistence agricultural systems; they also heightened competition for land.

Many of the plantations distributed to members of the political entourage of Mobutu remained under-utilized while large parts of the land formerly held under customary law were granted to indigenous rural capitalists or international agro-industrial companies. While these plantations offered new labour opportunities to the local population, many were either left vacant (with a loss of labour opportunities as a consequence) or guarded by a *gérant* (manager) who leased parts of the land to landless farmers. Land and food insecurity, and a labour surplus among the rural population, were among the most severe consequences. In the most highly populated regions, access to vacant land through the former customary contracts became impossible because customary held lands were no longer available. Buying land (which transferred control over land from the traditional authorities to the state and gave definitive land access to the new owner), however, was beyond the reach of most farmers due to a lack of financial resources. This forced most farmers to rent a tract of land under short-term contracts. In extreme cases, farmers were forced to work as agricultural labourers on a daily basis, often in return for extremely low wages.

The effects of shifts in land tenure patterns were reinforced by growing demographic pressure and consequences of the withering of the state. In the period 1958–1993, the population of Zaire soared from 15 to 42 million inhabitants, while production per capita shrank approximately 65 per cent, from $377 in 1957 to $117 in 1993 (Devey, 1997). In 1993, the total income of the Zairian state dropped to only $302 million. Agricultural production no longer generated state resources: in 1990, more than 80 per cent was for self-consumption and only 2.5 per cent for cash crops. In the eastern parts of the country, these conditions intensified local competition for land and reinforced the hardening of social boundaries, itself the result of increasing land scarcity. A survey in 1985 demonstrated that even with intensive cultivation, the land holdings of nearly 90 per cent of the population in Mulungu (Kabare, South Kivu) were insufficient to provide a family's minimal income. More than two-thirds of all households worked plots of less than 1 ha and one-third of all families had even less than 0.3 ha (Schoepf and Schoepf, 1987).

These dynamics explain why in Walungu, the tenure and use rights of non-customary land came to be increasingly regulated by alternative forms of contracts, such as individual ownership based on registered land titles. Purchasing land was almost impossible for most households, however, given their lack of financial resources. The only option remaining was to rent land, often under very insecure contracts: the use rights of the rented land were limited to (usually) one season, while only seasonal crops could be cultivated. When land was rented from churches or local landowners (village-level customary chiefs), the contracts usually included the additional obligation of providing one day of labour per week.[3] A final option was the payment of tribute in the form of labour. Before the war, farmers that did not have the means to rent land could work as labourers on plantations. However, beginning in 1998, production at most plantations came to a standstill.

In Lubero as well, land access became increasingly limited starting in the mid-1980s. There, two additional dynamics further reduced farmers' land rights and the available arable space. On the one hand, conflict between local customary authorities led to further erosion of smallholders' access rights as tenure rights became very insecure. In an area already densely populated, the competition between two customary chiefs led to the existence of two, parallel systems of governance, each of which committed itself to regulating central Lubero's land rights. An incomprehensible web of alliances and regulations totally blurred the distinction between state and non-state and between customary and civil regulation of ordinary people's access to arable land. There were numerous accounts of corruption in judiciary litigations, of people having to pay their tributes twice to different power holders, or of simply being chased from their land.[4] On the other hand, access to newly created plots in a former forest reserve was thwarted by the interference of Butembo businessmen, who profited from a governmental decree to buy large concessions for their cattle farms.

In order to deal with the increasing land pressure in Lubero, a governmental decree was issued on 31 January 1981 that granted local farmers access to part of the Itondi forest reserve of central Lubero. Two-thirds of the reserve was freed up for the local population to use for agriculture. However, in 1987 the liberated part of the Itondi forest was taken over entirely by a group of local businessmen from Butembo. Claiming the support of the local administration, they forced their way into the reserve and set up a number of large cattle farms, to the disadvantage of local farmers.

From the end of the 1980s, growing land scarcity provoked intensified tension between households. After Mobutu's announcement of a democratization process, this tension was easily manipulated by local elites in search of a popular base of support. As a consequence, local struggles for land access were transformed into ethnic animosity. In most cases, old claims of land ownership were exploited by local elites. The authority of customary authorities was rejected by members belonging to other communities on the argument that control over land was the right of the community that settled first. As a result of these dynamics, previous land allocations came to be judged on the criterion of ethnic citizenship (Van Acker, 2005) and existing structures of security came to be more rigidly defined on ethnic criteria.

One example of this is the events in Walikale, where land issues were manipulated by local elites, leading to violent clashes between different rural militias. When in the early 1990s poor Hutu Banyarwanda farmers living in Masisi lost their land because local customary chiefs sold it to rich absentee landlords of Banyarwanda origin, these farmers settled in Walikale (west of Masisi). Fearing the growing influence of the newly arrived Hutu Banyarwanda, both the local population and their Nyanga chiefs rejected their land claims. A coalition of local autochthonous political elites, afraid they would lose their political power if the Banyarwanda were registered as Zairian nationals and participated in the coming elections, started an exclusion campaign against the Banyarwanda. In Walikale, the campaign translated into increased animosity against the newly arrived Banyarwanda farmers. As a response to the position of the local elites, a local Hutu Banyarwanda association started encouraging its members to refuse to pay tribute to the autochthonous chiefs and to no longer recognize their authority. When indigenous customary authorities in Masisi also started to feel threatened by the Banyarwanda communities, the chiefs supported the formation of local militias. In March 1993, these militias started attacking Banyarwanda in the Masisi region. The result was a bloody confrontation that lasted for more than six months and killed between 6,000 and 10,000 people, while displacing more than 250,000 others.

The direct relationship between problematic access to land and conflict was further consolidated during the Congolese war. Local disputes over land came to be linked to the larger, multi-level conflict for political power and control over local resources. Land, however, also became an integral part of strategies by new coalitions – comprising local and regional actors – to acquire control over local economic assets and social mobility. The result of the new dynamic was that land gradually shifted from a *source* to a *resource* of conflict. The most visible illustration of this trend was the confiscation by local army commanders of local land titles. Access to land provided new politico-military leadership with the necessary economic basis, yet at the same time offered them a perfect resource to be distributed among their supporters. This was particularly the case in Masisi, where land was distributed to members of the new leaders' own ethnic groups, and other ethnic groups were forced to abandon their parcels.

A good illustration of this shift is the strategy developed by Banyarwanda elites in Masisi (North Kivu), for which land became a crucial element of a larger strategy to consolidate economic and political power. The campaigns by rebel groups against Kinshasa of 1996 (when Kabila started a campaign against Mobutu) and 1998 (when the RCD rebel movement tried to oust Kabila from power) had advantaged the Banyarwanda elites in their claims to land, which had become one of the crucial assets around which a new local power structure under control of the Banyarwanda elite has been constructed. The institution of local self-defence groups, originally meant to reduce the levels of insecurity caused by Interahamwe groups (Rwandan Hutu militias) and Mayi-Mayi militias (indigenous rural armed groups), was the first instrument to serve the political and economic interests of Banyarwanda elites. These self-defence groups were increasingly mobilized for the promotion of the interests of the Banyarwanda, both in military and economic terms. The Tout pour la Paix et le Développement (TPD), a non-governmental organization formed to facilitate the return of Banyarwanda refugees from Rwanda to North Kivu, was another instrument to strengthen control of Banyarwanda over local resources and livelihoods. This association was backed by important landowning Banyarwanda elites as an instrument to protect their economic interests. Through their influential positions within the RCD-ML rebel movement, which controlled the area between 1998 and 2003, Banyarwanda leaders also claimed control over the local administration. This power structure further stripped autochthonous customary chiefs of control over land. Local administrators and customary chiefs that did not support TPD were systematically replaced. In order to facilitate their access to land, Banyarwanda elites also campaigned to extract parts of Masisi from customary control and turn them into *zones extra-coutumières*, aimed at reducing the power position of the main indigenous ethnic group (the Hunde) that claimed full control over the local land tenure systems. For Hunde farmers, the customary chiefs were the protectors of their land rights and the only institution with the power to deal with land disputes. While the Banyarwanda rejected the exclusionary character of this power system, the Hunde defended the existing rural order on the premise of historical land rights. Since the start of transition in June 2003, however, most of the excluded Hunde chiefs were re-installed.

As these examples reveal, the role of land cannot be neglected when explaining local dynamics of conflict in eastern DRC. It is clear that access to land served a number of crucial functions. It provided a new local power structure with a necessary economic base yet at the same time helped to consolidate the support of certain parts of the grassroots population (based on ethnic affiliation), which in return was granted access to land. The informal governance structure that emerged from this unequal resource attribution – and which included both military and political elements – laid the foundation for further reinforcement of ethnic boundaries. In eastern DRC, these dynamics pushed farmers belonging to other ethnic communities further into a very insecure economic position. Reduced access to land and considerable levels of insecurity strongly limited agricultural activities and seriously affected existing food systems. This was particularly the case where an environment of continuing insecurity continued to diminish economic alternatives. Insecure land access, decreasing agricultural production and the disruption of rural markets forced poorer households to develop alternative strategies of survival. In the next section, which is based on fieldwork in Walungu, Masisi and

Lubero, we analyse a number of these strategies, as well as their impact on the local food security situation.

Effects of land scarcity on food security and rural livelihoods

The examples in the previous section demonstrated that the Congolese conflict not only further reduced land access but also disrupted existing mechanisms for protecting land property, thus eroding the legal position of rural households. The interference of rebel administrations in customary systems of land distribution also complicated the land tenure insecurity of most farmers. A lack of legal protection facilitated the acquisition of land by local elites and rebel leadership structures, while the absence of a clear legal framework of local land distribution allowed the rebel leadership to create institutional mechanisms to distribute land titles to its own supporters. One of these mechanisms was the Commission de Lotissement in Masisi centre, created in 2003 and composed of several provincial services, which gave advantage to supporters of the politico-military leaderships in their search for land. Military commanders and local administrators also interfered increasingly in the distribution of land, often to their own benefit. In Masisi, local defence forces and Mayi-Mayi militias granted farmers access to land in return for payments for the war effort (the confiscation of harvests was a widespread practice) and evicted by force farmers who refused to pay those 'taxes'. Local administrators, such as agronomists and representatives of the provincial administration, sold plots illegally to local farmers. According to local sources in Mahele (Masisi), an entire locality was sold illegally by administrators of the provincial Services de Cadastre et Titres Fonciers. In many cases, these transactions were based on ethnic criteria. Corruption and poor functioning of local courts increased local farmers' land tenure insecurity, as it became extremely difficult to claim legal property rights.

The lack of legal protection, growing land scarcity, the generalized context of insecurity and the redistribution campaigns by rebel commanders placed growing pressure on local farmers' access to food. The most direct consequence of these dynamics was a sharp reduction in food production. According to one survey, the agricultural production in Walungu decreased by 75 per cent between 1996 and 2005 (OCHA, 2005). Eventually most local production was reserved for self-consumption. A particular effect of the war was the disappearance of cattle, which not only influenced local food production but also had a number of effects on local social cohesion as it disrupted several social traditions. In Walungu, as in many other places, prior to the war cattle were not only an important economic asset but also regulated social relations. Traditionally, cattle, usually kept on communal grounds on the top of nearby hills, were one of the central elements for land tenure contracts, as they were the rents to be paid when land access was granted. They also helped households pay for medical services and school fees, were an important component of the local diet (milk and meat) and were used for fertilizing land. In addition, cattle

Box 10.2 Displacement and land access in Masisi

In Masisi during the war, land access was complicated by new claims of returning refugees and Banyarwanda IDPs, or by IDPs who wanted to settle permanently in the areas where they had been living since the war. On their return from exile, many farmers who sold their lands before leaving their villages (often at a very low price) tried to renegotiate their former land properties, which in many cases resulted in disputes between former and present owners. In some cases, inter-ethnic committees of elders helped reach agreements locally. For example, the elders in the village of Burungu assisted former landowners (who had been forced to sell their land because of mounting insecurity) to renegotiate the deeds of sale; they had their land returned and refunded the money paid. In other areas, however, land that had been sold was regained by the use of force. In Kibabi, returning Banyarwanda refugees were accompanied by armed elements who drove the new owners from their land. Several reports claimed that the provincial administration and local rebel leadership (through TPD structures) were involved, but there is little evidence about the extent and aims of the process. Another source of tension in Masisi was the presence of squatters on plantations and cattle ranches. From 1999, many returned refugees and IDPs were given access to cattle ranches and large farmlands to cultivate as squatters. In other cases, land was rented from large landowners, often at very high prices and leaving only limited profit margins to the farmers. Another practice was the granting of land to households on the basis of sharecropping, which included a claim of part of the crop in return. These patterns of land access were very insecure. In cases where households were not able to pay the rent, they were often forcibly evicted by the landowners. Squatters granted access through a local gérant to land owned by people who went into exile during the 1990s as a result of high levels of insecurity were evicted from the land when their original owners returned. In some cases, former rebel soldiers served as private security forces for wealthy and politically connected landowners.

served as an important safety net for households in cases of crisis and were the material basis of marriage contracts. During the war, however, armed groups pillaged or killed most of the cattle. Communal farms became inaccessible due to high levels of insecurity, and the treatment of cattle diseases was disrupted due to lack of technical support and financial resources.[5]

Other elements that caused a decrease in local food production were related to environmental factors (erosion, increasing infertility of land and crop diseases) and the disappearance of technical support systems and limited access to seeds and tools. One illustration is the case of Lubero, where insecure land access forced peasants to precipitate the sowing of their fields for fear of leaving them fallow and otherwise exposing them to seizure by landlords. Lack of respect for the agricultural calendar exposes crops to climatic risks and hence negatively influences agricultural output. Another environmental factor in the region was the severe erosion of arable space. Even if not a direct consequence of the Congolese war and already being observed prior to the conflict, peasants were increasingly reticent to plant anti-erosive hedges in part because it decreased cultivable spaces and invited rodents that devastate crops.[6] Scarcity of land also made peasants less likely to adopt agricultural practices recommended by aid organizations. Fields were overexploited, which reduced the fertility of land.

The effects of the collapse of food production were reinforced by the reduction of commercial activities. Before the war, local networks of markets connected remote areas with the provincial capitals Bukavu and Goma through a number of local markets. Walungu was an important provider of manioc to Bukavu, while Masisi provided Goma and Kinshasa with meat and several agricultural products. During the war, most of these markets were moved or witnessed a sharp decrease in activities. Insecurity, lack of local production, increased prices and limited purchasing power all gave rise to the disruption of commercial activities. The disappearance of credit systems also disturbed trading activities. Before the war, businessmen from Bukavu granted credit to local traders in Walungu to facilitate the purchase of local agricultural products. When those credits were no longer offered, small traders lacked the financial means to buy products. A good indication of this loss of purchasing power was the reduction in the size of standard measures used at markets for the sale of agricultural products.

Food prices were often imposed by businessmen from the main cities, while the absence of cooperatives and associations meant that petty traders lacked the capacity to protect their interests. A final constraint to trading activities was the introduction of additional taxes by rebel groups and the confiscation of purchased goods by armed elements. In order to escape rising taxes, a number of small markets were introduced at the village level, where goods were bartered. These elements explain why markets in Walungu no longer connect South Kivu's hinterland and mining centres to Bukavu. Prior to the war, those markets traded minerals and local agricultural products in return for manufactured products, salt and soap.

In the various research regions, lack of food production and limited access to markets resulted in generalized food insecurity. One survey concluded that in Walungu, 72 per cent of households had a monthly income of less than $30 and 22 per cent had an income of between $30 and $50 (Bahirire, 2004). Another study based on a survey of 840 households in a number of locations of South Kivu, including Walungu, showed the shifts in the number of meals per day, offering a clear indication of the food security situation (Diobass, 2005) (see Table 10.1).

Table 10.1 Number of meals consumed per day in South Kivu

	Households			*Children between 0 and 5 years*		
	Before war	*During war*	*In 2004*	*Before war*	*During war*	*In 2004*
<1 meal/day	0%	3.4%	21.9%	13.5%	30%	14.6%
1 meal/day	0%	83.3%	21.6%	0%	30%	16.6%
2 meals/day	40%	10%	44.9%	34.4%	30%	51.1%
3 meals/day	60%	3.3%	11.6%	52.1%	10%	17.7%

Source: Diobass (2005)

From coping to adaptation: Shifts in local food systems

The previous sections showed that the Congolese conflict and consequent shifts in the institutional environment regulating access to land not only provoked more and intensified struggles for land. They also put existing food systems under stress, transforming them from predictable mechanisms of production, processing, distribution and consumption into very volatile, unpredictable and uncontrollable systems of survival (Pingali et al, 2005). Chronic violence and conflict tend not only to reshape existing mechanisms of food production, and to distort the available assets and choices of households, but also to provoke the same households into a 'struggle to adapt their livelihoods systems to accommodate violence' (Lautze and Raven-Roberts, 2006). When food insecurity is prolonged, as was the case in eastern DRC, damage to livelihoods becomes inevitable, assets are turned into liabilities and people and households are forced to adopt alternative strategies to survive. The question arises, then, to what extent are these longer-term new strategies parameters for measuring changes in livelihoods systems in general and food systems in particular? In other words, short-term coping strategies might indicate the first phase of a larger process that includes the development of longer-term strategies for adapting to crisis contexts. A next question relates to the outcomes of these strategies. Do the different adaptive strategies of rural households result in greater resilience of their livelihoods and a move towards food security, or do they lead to increased vulnerability to external shocks? As was also seen in the chapter on Beni-Lubero (Raeymaekers, this volume), household attempts to make their livelihoods less vulnerable to violence and protracted food crisis do not necessarily make them more resilient. This section analyses a number of short-term and long-term strategies that have been developed by rural households in Walungu, Masisi and Lubero to cope with the impact of declining food entitlements and land scarcity, and evaluates to what extent these strategies affected land security.

In the regions under study, increased land scarcity, high levels of insecurity and the absence of customary protection mechanisms forced many households to develop alternative strategies in order to guarantee their livelihoods and food security. These strategies produced a number of shifts in local food security mechanisms and drastically changed food production patterns. As fieldwork in Masisi demonstrated, during the war agricultural production was mainly reserved for self-consumption, with a significant shift from extensive to intensive cultivation and from perennial crops to low-risk and seasonal crops. Traditionally the main crops in Masisi were beans, bananas, sweet potatoes, green peas and maize. Today they are beans, soybeans, sweet potatoes and maize, with manioc and peanuts in the lower parts of Masisi. These shifts in crops indicate that agricultural production was increasingly guided by the minimizing of risk rather than the maximizing of profits. In addition, the diversification of crops was in accordance with tenure security, which explained the reduction of perennial crops. Most households, however, have also employed a number of strategies to reduce their dependency on land access and safeguard their food security, including migration to the urban centres and mining sites and the development of small commercial activities.

In Walungu, fieldwork revealed a similar trend. There, the production of cash crops such as bananas, coffee, tea and quinquina had disappeared almost entirely. The shifts in crop types indicated that agricultural production had become driven more by the push to minimize risk than to maximize profits (crop diversification and perennial crops require tenure security). The reduction in the size of cultivated plots had also led to a shift from monoculture to polyculture of crops: on small plots, crops that could be cultivated in the short term (sweet potatoes and beans) were planted in combination with manioc. Such polyculture had a devastating impact on the quality of the land (especially because the land was hardly fertilized), but was necessary to guarantee the production needed for self-consumption and reduce the risk of loss to looting. In many cases, the crops were also harvested prematurely in order to further reduce the risk of pillaging. In Walungu it was reported that some landless households stole food from other households' fields. Another strategy was to extend the size of plots by confiscating part of the neighbouring plots, which became a major source of disputes between farmers on the borders of their parcels. A final strategy observed was the exploitation and transformation of wetlands into small plots. In Bikeka (Walungu), by 2005 an estimated 75 per cent of households relied on such plots for their survival.[7] They were producing vegetables such as onions, cauliflower and tomatoes, which in most cases were sold at local markets in order to pay for manioc (before the war, Walungu was a region that provided manioc). Under the customary system, wetlands could be cultivated only to deal with shocks and food insecurity; as a consequence of the war they tended to be overexploited.

In Lubero, it was observed that since the war and the consequent reduction of land access, peasants were no longer allowing their fields to lie fallow. Several reasons explain this strategy. When no crops were present in a field, customary landlords automatically concluded that it was unexploited, and, eager to increase their interests or annual payments, subsequently leased it to another peasant. A non-cultivated field also ran the risk of being grabbed up by competing farmers. On numerous occasions members of the same family initiated physical fights over access to plots: while certain family members believed themselves to hold the rightful title to a field, other members arrived at night to sow it in secret, with the obvious result of diminished productivity. As a consequence, some households started to organize sentries to protect their fields from invaders. These sentries usually involved a group of three to five males who camped in the field at night to ward off trespassers.[8] In Lubero, another consequence of the increasing pressure on land was that farmers were forced to cultivate in more remote areas. Traditionally the arable space in Lubero's highlands was concentrated in the vicinity of villages. Because of mounting population and geographical pressures, however, peasants were forced to cultivate in more distant areas. This often meant several hours of walking between the village and the field. The widening of this distance led in turn to a spatial division between vegetable gardening (cultivated in the vicinity of the home) and monocultures (in the more distant fields). Another strategy was to move to less populated areas in the western forest belt, which had some advantages in terms of immediate availability of land and agricultural income. The total lack of infrastructure and services in the area, though, meant the efforts of those dislocated households were largely in vain.

Other strategies to deal with the effects of limited food production aimed at making food security less dependent on land access. However, these strategies were hardly able to produce greater resilience in relation to food security. The development of commercial activities and migration to urban and mining centres became a very popular alternative to agricultural production. While migration to the mining centres was not a direct consequence of the war, beginning in 1996 an increasing number of men opted for that strategy in search of economic opportunities, leaving their families behind. Others moved to Bukavu or local population centres in order to escape from daily acts of violence or joined local militia groups and themselves became a source of insecurity. More important even was the shift in main economic activity from agriculture to trade. As one observer in Mugogo (Walungu) stated, everyone became a trader. Women in particular tended to invest in small commercial activities and trade such products as manioc flour, salt, palm oil and fish, although the income generated from those trading activities remained extremely low due to lack of purchasing power and market opportunities. Rather than producing food and selling it at local markets, however, much of the food sold in the region was purchased in Bukavu. Because by late 2005 traditional centres of manioc production such as Kalonge, Tubimbi and Mulamba were still under control of FDLR groups (Rwandan Hutu rebels)[9] and elsewhere production was limited by mosaic disease, locally consumed manioc was imported from Bukavu, which caused its price at local markets to increase considerably. Traditional systems of solidarity that used to regulate trading activities and protected small traders, such as not buying from the same producer twice, were no longer practised.[10]

Finally, limited land access and a context of generalized insecurity also affected local food patterns. In Walungu, a traditional meal consisted of manioc, milk, meat and vegetables. By 2005, the basic meal was just manioc. Due to the disappearance of cattle, meat and milk were no longer available and had been replaced by products that traditionally were prohibited or reserved for only some parts of the population. One example was the increased consumption of chicken, which according to local tradition could not be eaten by women. Another example were the *cobay* (guinea pigs), which traditionally were reserved for children but came to be consumed by all members of society, even at ceremonies. Rice and maize also came to be part of the local food consumption patterns as a result of food security interventions by international agencies and local associations.

Food security interventions[11]

Over the last few years concerns about the impact of policy interventions on food entitlements of crisis-affected people and households have provoked an intense debate among researchers and agencies on how to build more effective policy responses. This debate has revealed a number of common deficiencies of policy responses to food insecurity. First, it was observed that the international community had moved from long-term, development-oriented assistance to short-term emergency support. Secondly, while protracted crises are to be understood as complex processes that are deeply rooted in local society, interventions in most cases tended to start from a top-down and blueprint approach and were limited to a standardized set of responses. Moreover, most donor agencies regarded conflicts as an aberration from the normal path of development and neglected the complexity and changing environment of protracted crises (Korf and Bauer, 2002). The main reason for this seems to have been the lack of assessments of livelihoods as well as a weak link between available information about the crisis environment and policy formulation. Thirdly, these short-term interventionist frameworks tended to be commodity-focussed and dominated by food aid or the provision of seeds and tools, thus neglecting the other dimensions of food security (Flores et al, 2005).

The case of eastern DRC reveals that while crisis-affected households try to cope with the generalized conditions of violence and to adapt their livelihoods systems to the conflict-environment, most food security interventions also tend to neglect these coping strategies and the implications of the institutional shifts on the food security situation. Even if most international agencies recognize the need to support livelihoods systems in conflict situations, they often suffer from a 'livelihoods gap' (Le Sage and Majid, 2002): standard relief interventions tend to focus on direct access to food and other relief supplies, and fail to address 'the complex web of spatial and temporal vulnerabilities generated by… violent conflict' (Lautze and Raven-Roberts, 2006). This is clearly illustrated in the different food security interventions in eastern DRC. While most interventions focused on direct access to food through the distribution of relief supplies, in very few cases were those short-term interventions based on assessments of the food security situation and structural or institutional causes of inequality and food insecurity.

From the start of the Congolese conflict, several international agencies and local associations addressed the direct food security needs of the rural population. Very few agencies, however, dealt with land issues. Most of the food security interventions tended to focus on short- and medium-term food needs, including food aid, the support of nutrition centres and the distribution of seeds and tools. In Walungu, alternative crops (such as soya and maize) were introduced that aimed at increasing the local production of food and the improvement of the nutritious status of the local population. While most local households were originally hesitant to consume the crops (a large amount of the distributed maize was resold at local markets), the interventions eventually resulted in shifts in local food consumption patterns and an improvement in the nutrition status of the population.

As in central Lubero, however, most interventions were based on a standardized set of responses that did not take into account the dynamic nature of its food system. Due in part to a lack of participatory planning methods, most concentrated narrowly on food production, without taking into consideration other dimensions of food security that proved to be significant for peasants. While the interventions led to a relative increase in food production because of the increased availability of seeds, they were insufficient for addressing in a sustainable way the underlying problems of the food crisis.[12]

Beginning in 2002 an increasing number of international agencies expanded their focus from the introduction of mechanisms to reintegrate IDPs and refugees, and reforestation and anti-erosion measures, to the improvement of rural infrastructure and road networks. In Masisi, the rehabilitation of the local road system by Agro-Action Allemande generated a number of positive effects. One strategy was the improvement of the road networks through food-for-work programmes, which also aimed at assisting chronically undernourished people. While there were significant drawbacks to this food for work approach,[13] in the case of Masisi the rehabilitation of roads improved the local security conditions considerably and had a positive effect on the food security situation. Better and safer roads facilitated local access to food by increasing economic exchanges and the return of IDPs and refugees. In Masisi, the overall effect of road rehabilitation led to an estimated 50 per cent growth in agricultural production.

In the various regions, local institutions and actors played a limited role in humanitarian responses. Most interventions were organized by external aid agencies, with the role of local actors limited mostly to the execution of agency programmes; they were planned according to a top-down decision-making process and were not based on local responses to food insecurity. As a result, local food economy structures and long-term programmes developed by local associations were often disrupted. The free and unconditioned distribution of food, for example, disturbed local support programmes based on the distribution of microcredit. However, in each region, local community-based structures and humanitarian organizations gathered significant knowledge and expertise about local food security mechanisms and developed their own mechanisms and strategies to mitigate the effects of food crises.

In Walungu, which is characterized by very strong associations, an increasing number of local actors tried to address the local food insecurity situation. One strategy was the distribution of seeds through seed rotation systems. Under this system, farmers who received crop seeds had to donate part of their production to the seed distributing association, to be redistributed to other members of the association. Other associations introduced *greniers de semence* (seed barns), which required farmers to give back after harvest the same amount of seeds they had received; the barns guaranteed the seeds necessary for the following season. In most cases, support to farmers was not limited to seed distribution but included technical assistance as well. Other interventions aimed to reintroduce livestock through rotation systems or reintroduced traditional livestock management systems (*systèmes d'élevage de stabilisation*). The main objective of these interventions was to improve the quality of land and food production.

Even if problematic access to land has been recognized as one of the main reasons for food insecurity and local tension, very few humanitarian agencies integrated a focus on land into their humanitarian interventions. An interesting exception was the introduction of *chambres de pacification* or *chambres de paix* (pacification councils or peace councils) in Walungu. A number of local associations introduced mechanisms to strengthen the capacities of local farmers to claim their land rights and to help resolve disputes over land. Local councils (*chambres*) composed of elders investigated the nature of the dispute and tried to reach a solution based on a compromise between the farmers involved in the case. While these mechanisms served to further informalize justice, they were the best mechanisms available for offering some protection to local farmers. Farmers mistrusted customary justice systems and local courts because of corruption (in the courts, the one who paid most usually won the case) and lack of legal protection. The main disadvantage of the informal *chambres* was that access was usually limited to the members of the association that managed them. That condition tended to exclude non-members from legal protection; this was especially true for church-based organizations. Most of the associations that introduced conflict resolution procedures also informed farmers about their property rights by distributing information on the legal framework regulating access to land; many developed advocacy efforts at the national level to modify the existing land laws.

A similar approach was used in Lubero, where the local SYDIP worked to reinforce the legal position of local farmers through the introduction of a team of peasant lawyers that mediated between opposing parties of peasants involved in land conflicts. When no amicable solution was reached, farmers were assisted at the judiciary level. Although still in its early stages, the mediation between conflicting parties had already resulted in a modest reduction in corruption on the part of the judges, and increased respect for the judiciary's final decisions.

In Masisi, there were very few programmes aimed at strengthening the legal position of local farmers. One notable exception was the programme of Aide et Action pour la Paix (AAP), a local organization that in 2004 distributed background information about the existing land law to other local associations in order to assist in the resolution of land disputes in Masisi. Local conflict resolution mechanisms like those in Walungu, however, remained absent in Masisi. Most interventions were limited to short-term food aid and medium-term interventions to reinforce local food production but did not deal with the structural causes of food insecurity. This could be explained in part by the fact that as a result of the war, land issues had become completely politicized and disputes over land access between households were dealt with in most cases by local army commanders. Many interventions also suffered from a lack of reliable information on the food security situation in the more remote areas. The only studies focused on eastern Masisi, where Save the Children-UK and the local organization Asrames had conducted a number of surveys on the food security situation and household economy, while Save the Children-UK, Mèdecins sans frontières and World Vision International, among others, had conducted nutrition surveys. On land issues, the only study conducted was a livelihood analysis done by Save the Children-UK in 1999.

Table 10.2 summarizes the constraints to food production and access, household responses and the interventions developed by food aid agencies and local associations.

Conclusions

This chapter analysed the problematic relationship between land tenure, conflict and food security in eastern DRC. It explored how limited land access and growing levels of insecurity influenced household strategies to acquire access to food and how food security interventions tried to deal with increased lack of food access. Finally, it evaluated to what extent interventions had addressed land tenure insecurity, which was identified as the main structural cause of food security. The analysis was based on extensive fieldwork in three regions: Walungu, Masisi and Lubero. The areas were characterized by problematic land access and had recently witnessed violent conflict, which was either caused by intensified competition for land or reinforced local struggles for land.

Each of these cases revealed that problematic land access, at different levels, had been a key dynamic of local tension and conflict. In Walungu, Masisi and Lubero, reduced access was mainly the result of the institutional context regulating the distribution of land. Colonial rule and patrimonial practices under Mobutu eroded existing customary institutions and patterns of land distribution, and facilitated the land-grabbing by a new class of rural capitalists, to the disadvantage of rural households faced with insecure access to local assets and loss of their land. In all three areas, population growth intensified these effects and provoked greater competition for land between individual farmers. However, insecurity of tenure and demographic pressure did not necessarily lead to violent conflict or turn local disputes into increased political antagonism. A key condition seemed to be the presence of different ethnic communities within the same area and the attempts of local entrepreneurs to mobilize and pitch ethnic communities against each other.

An often-neglected dynamic in the analysis of linkages between insecure land tenure and conflict is the use of land as a 'resource' of conflict. The case of Masisi revealed that land can be turned into a foundation for local power bases and an asset for former rebel leaders and politico-military elites to reward their political supporters. These practices in turn can intensify local competition

Table 10.2 Food constraints, household strategies and interventions

Constraints to food production	*Constraints to food access*	*Household strategies*	*Main food security interventions*
Institutional factors: • Shifts in land tenure systems • Limited access to land • Lack of technical support by local agronomists • Crop restrictions on rental land • Reduction of plot sizes • Non-respect of customary titles and lack of legal and customary protection • Conflicts over land ownership between former and new owners • Restricted access to plantations • Disappearance of cash crops • Limited availability of seeds and tools • Loss of livestock ***Environmental conditions:*** • Erosion • Crop diseases • Decrease of land fertility due to over-exploitation ***General conditions:*** • Demographic pressure • Insecurity caused by armed elements • Pillaging of harvests • Increased pressure on land because of the presence of IDPs	• Reduction of purchasing power • Limited availability of food • Increase in the amount and number of taxes • Limited access to markets due to insecurity • Absence of cooperatives that protect small traders • Lack of organization • Pillaging of stocks • Degradation of road network • Destruction of market infrastructures • Disappearance of microcredit systems	• Rental of land under bwasa contracts • Cultivation of small plots in wetlands • Shift from monoculture to polyculture • Adaptation of crop diversification to (lack of) tenure security • Shift from agriculture to petty trade • Shifts in food consumption patterns • Harvest and consumption of immature crops • Cash for work • Joining of farmers associations • Cultivation on shared plots • Migration to urban centres or mining sites • Joining of local militias • Theft of crops	• Food aid distribution (by international agencies, often through local associations) • Distribution of seeds and tools, introduction of greniers de semence (by international agencies and local associations) • Provision of technical support (by local associations) • Introduction of collective fields (by local associations) • Rehabilitation of rural infrastructure (roads etc) (by international agencies) • Introduction of livestock rotation mechanisms (by local associations) • Institution of conflict resolution mechanisms (by local associations) • Introduction of micro-credit systems (by international agencies)

Source: Author compilation

for land and engender new dynamics of conflict. The studies in Walungu and Lubero, however, demonstrate that loss of land and demographic pressure do not automatically induce violent conflict, although at a grassroots level land disputes do undermine local social cohesion. It is only once local political or military leaders begin using the land access problem in their efforts to mobilize grassroots populations that land issues are translated into more intense conflict.

Since the start of the Congolese war in 1996, the harmful effects of increased land alienation intensified in every case we saw. Generalized conditions of insecurity forced farmers to migrate to safer regions and had a devastating impact on the local production of food. Conflict also limited traditional livelihood resources, seriously damaged the remaining infrastructure and disrupted local markets. The protracted humanitarian crisis further increased pressure on land and disrupted mechanisms of land use protection, while the disappearance of cash crops led to considerable reductions in household income. On an institutional level, the customary chiefs' loss of power and control over land (which itself had been a direct outcome of colonialism and the capitalization of land under Mobutu) was further reinforced during the war, to the advantage of politico-military leaderships.

A good illustration of this dynamic was the situation in Masisi, where customary protection mechanisms were replaced by new mechanisms of private protection under control of political elites and military commanders. Customary land titles became less secure and large parts of the land were (illegally) granted to local businessmen and rebel leaders, leading to increased land tenure insecurity at a local household level. As a result of shifts in the local politico-military equilibrium, in Masisi land came to be viewed as a resource of conflict, to be distributed mainly to the supporters of the new leaderships. At the same time Banyarwanda leaderships tried to seize upon the opportunity to advance Banyarwanda goals of self-determination for the area, including with regard to land (Unruh, 2003). This led to the development of multiple, informal normative orders regarding land tenure and access.

In each case studied, household responses to the effects of land scarcity and the Congolese war engendered a number of shifts in local food systems and consumption patterns, yet these did not necessarily lead to greater resilience of their livelihoods nor of their food security. The most visible impact was the modification of local food production arrangements. Land tenure insecurity provoked a shift from perennial crops to seasonal crops and from monoculture to polyculture of plots. It led to the (near-) disappearance of cash crops and turned unexploited wetlands into small agricultural plots. Peasants started cultivating in more distant areas, which led to a spatial division between vegetable gardening (cultivated in the vicinity of the home) and monocultures (on the more distant fields), or to migration to less populated areas. Farmers who migrated to other regions faced significant constraints to integrating into local social structures, which continued to be based on customary systems of land tenure. Local food consumption patterns also changed. As a consequence of the pillaging of livestock and a fall in food production, the number of meals consumed per day decreased and the daily diet was reduced to manioc and a limited portion of vegetables. In Walungu, meat and milk were replaced by products that were traditionally prohibited or reserved only for some parts of the population. Other products, such as rice and maize, were consumed mainly as a result of food security interventions by international agencies and local associations.

Other household strategies sought to reduce dependency on land access. Migration to the urban centres and mining sites became a popular exit strategy, especially for young men, and enrolment in local militias or the army was often viewed as an alternative for the younger generation. Women tended to invest in small commercial activities and to trade such products as manioc flour, salt, palm oil or fish. The income generated from the trading activities was extremely low due to lack of purchasing power and market opportunities (see Table 10.2).

In general terms, however, it should be recognized that in the areas under study household responses to the local food crisis, which included mainly short-term coping strategies, did not lead to greater resilience of rural livelihoods but on the contrary tended to reconfirm the structures and processes that had caused food insecurity. In their pursuit of survival mechanisms, households lacked the capacity to make an impact on the institutional context that defined their access to crucial resources, or to adapt to the conditions that were causing limited access to these assets. In the areas studied, land tenure security was the most crucial asset in determining local food security conditions. Access to this asset, however, was increasingly limited during the war or monopolized by local politico-military elites who used it to reward supporters and further eroded existing land tenure institutions. In most cases, households were not able to deal with the changes and remained dependent on a series of exploitative institutions and networks.

International agencies and local associations have tried to address the generalized conditions of food insecurity in eastern DRC but have neglected the institutional context defining people's production of, and access to, food. A review of the main interventions in each case revealed that interventions had focused mainly on short-term, direct access to food through the provision of food aid, seeds and tools distribution, and on the establishment of nutrition centres. The interventions often neglected the potential role of local associations. In other cases local efforts to deal with food insecurity (such as seed distribution systems) were disrupted by external interventions. The shifts in local food consumption patterns caused by food security programmes also indicated that international donors do not fully take into account the actual and potential roles of local associations and institutions and that short-term interventions are not fully integrated into longer-term policies. Finally, few interventions were based on assessments of the food security situation. The lack of reliable and comparable information, which is a result of the inexistence of local information systems and the near collapse of local administrative structures, had reduced the impact of humanitarian interventions. However, little progress had been made in establishing valuable information flows. As a consequence, shifts in local food systems and institutional changes instigated by the protracted crisis and by household strategies of adaptation were seldom integrated into food security interventions.

Longer-term interventions barely addressed issues linked to the structural causes of food insecurity. In Masisi, Walungu and Lubero, most interventions neglected the issues of insecure land tenure and increasing land scarcity, even though these constituted the main structural factors of food insecurity in those areas. A limited number of local associations had tried to tackle the effects of land scarcity through the introduction of collective fields, the establishment of microcredit systems, the provision of judicial support and the institution of conflict resolution mechanisms. Others had focused on effects of land scarcity such as shifts in production patterns, increased impact of climatic conditions, loss of livestock and rehabilitation of local infrastructure. The impact of these interventions were limited, however, by the associations' lack of technical and financial capacities.

The case of eastern DRC confirms that food security policies and strategic frameworks for protracted crisis situations need to strengthen people's access to vital assets. This includes the reinforcement of protection mechanisms and of households' capacities to claim their land rights. The case of eastern DRC clearly shows that secure land access and control is a crucial condition for social reconciliation and conflict resolution. In rural areas, land is not only a vital economic asset, it also gives people an identity and offers them protection. Therefore food security interventions should be oriented towards longer-term issues of access and accessibility rather than short-term food availability. This includes a focus on the institutional conditions that determine people's access to local assets, such as land.

One constructive intervention could be to strengthen institutions, such as the *chambres de paix* in Walungu, for mediating conflicting interests of farmers and others with respect to local land rights. Another strategy could be to design all interventions by way of a conflict-sensitive planning process, in order to ensure that potential negative impacts are avoided or mitigated, particularly for programmes that might influence people's land use patterns. In order to be effective in protracted crisis situations, such a process would also have to include modifying conventional policy frameworks with respect to local conflict dynamics and adaptations. It would also require establishing information networks that provide the necessary data on the food security context and on the institutional environment that conditions people's access to food, to guarantee adequate analysis of the context of intervention. These mechanisms should be the result of joint efforts by donors and local development associations.

References

Bahirire, C. (2004) [untitled] (mimeo)

Broegaard, R. J. (2005) 'Land tenure insecurity and inequality in Nicaragua', *Development and Change* 36 (5): 845–864.

DeBacker, M. (1958) 'Etude sur le problème de reboisement sur des regions montagneuses autour du Graben africain', *Bulletin de l'Information de l'INEAC* 7 (3).

Deininger, K. (2003) *Land Policies for Growth and Poverty Reduction*, World Bank, Washington DC, and Oxford University Press, Oxford, UK.

Devey, M. (1997) 'L'économie Zairoise: état des lieux', *Marches Tropicaux et Mediterraneens* 2670: 57–96.

Diobass (2005) [untitled study by local organization] (mimeo)

Flores, M., Khwaja, Y. and White, P. (2005) 'Food security in protracted crises: Building more effective policy frameworks', *Disasters* 29 (S1): 25–51.

Hecq, J. and Lefèbvre, A. (1959) 'Eléments de la production agricole au Bushi: Recherche de la superficie nécessaire par famille', *Bulletin Agricole du Congo Belge* 9 (4): 285–293.

Huggins, C. and Clover, J. (2005) 'Introduction'. In C. Huggins and J. Clover (eds) *From the ground up. Land rights, conflict and peace in sub-Saharan Africa*, pp1–24, Institute for Security Studies (ISS), Pretoria.

Jackson, S. (2003) 'Fortunes of war: The coltan trade in the Kivus'. In S. Collinson (ed.) *Power, Livelihoods and Conflict: Case-Studies in Political Economy Analysis for Humanitarian Action*, pp. 21–36, HPG Report no. 13, Overseas Development Institute (ODI), London.

Korf, B. and Bauer, E. (2002) *Food Security in the Context of Crisis and Conflict: Beyond Continuum Thinking*, Gatekeeper Series No. 106, London.

Lautze, S. and Raven-Roberts, A. (2006) 'Violence and complex humanitarian emergencies: Implications for livelihoods models', *Disasters* 30 (4): 383–401.

Le Sage, A. and Majid, N. (2002) 'The livelihoods gap: Responding to the economic dynamics of vulnerability in Somalia', *Disasters* 26 (1): 10–27.

Mararo, B. (1997) 'Land, power and ethnic conflict in Masisi, 1940s–1994', *The International Journal of African Historical Studies* 30 (3): 503–537.

Maxwell, D. and Wiebe, K. (1999) 'Land tenure and food security: Exploring dynamic linkages', *Development and Change* 30 (4): 825–849.

Maxwell, D., Watkins B, Wheeler R. and Collins G. (2003) *The Coping Strategies Index: A Tool for Rapidly Measuring Food Security and the Impact of Food Aid Programmes in Emergencies*, paper presented at the FAO International Workshop 'Food Security in Complex Emergencies: Building Policy Frameworks to Address Longer-Term Programming Challenges', Tivoli, 23–25 September 2003, FAO, Rome.

OCHA (United Nations Office for the Coordination of Humanitarian Affairs) (2005) [untitled] (mimeo).

Pingali, P., Alinovi, L. and Sutton, J. (2005) 'Food security in complex emergencies: enhancing food system resilience', *Disasters* 29 (S1): 5–24.

Pottier, J. (2003) *Emergency in Ituri, DRC: Political Complexity, Land and Other Challenges in Restoring Food Security*, paper presented at the FAO International Workshop 'Food Security in Complex Emergencies: Building

Policy Frameworks to Address Longer-Term Programming Challenges', Tivoli, 23–25 September 2003, FAO, Rome.

Pottier, J. (2006) 'Roadblock ethnography: Negotiating humanitarian access in Ituri, Eastern DR Congo, 1999–2004', *Africa* 76 (2): 151–79.

Pottier, J. and Fairhead, J. (1991) 'Post-famine recovery in highland Bwisha, Zaire: 1984 in its context', *Africa*, 61 (4), pp. 437-470.

Schoepf, B. G. and Schoepf, C. (1987) 'Food crisis and agrarian change in the eastern highlands of Zaire', *Urban Anthropology* 16 (1): 5–37.

Sosne, E. (1979) 'Colonial peasantization and contemporary underdevelopment: a view from a Kivu village'. In G. Gran (ed.), *Zaire: The Political Economy of Underdevelopment*, pp. 189–211, Praeger, New York.

Tsongo, M. (1994) *Problématique d'accès à la terre dans les systèmes d'exploitation agricole des régions montagneuses du Nord Kivu*, Université Catholique de Louvain (UCL), Louvain-la-Neuve, Belgium.

United Nations Security Council (2001) *Report of the Panel of Experts on the Illegal Exploitation of Natural Resources and Other Forms of Wealth of the Democratic Republic of the Congo*, document S/2001/357 (12 April).

Unruh, J. D. (2003) 'Land tenure and legal pluralism in the peace process', *Peace and Change* 28 (3): 352–377.

Van Acker, F. (2000) *Of Clubs and Conflicts: The Dissolvent Power of Social Capital in Kivu (DR Congo),* Research Paper 2000-07, Faculty of Applied Economics, Antwerp.

Van Acker, F. (2005) 'Where did all the land go? Enclosure and social struggle in Kivu (DR Congo)', *Review of African Political Economy* 103: 79–98.

Vlassenroot, K. and Huggins, C. (2005) 'Land, migration and conflict in eastern DRC'. In C. Huggins and J. Clover (eds) *From the Ground Up. Land Rights, Conflict and Peace in Sub-Saharan Africa*, pp. 115–194, ISS, Pretoria.

CHAPTER 11

Beyond the blueprint: Implications for food security analysis and policy responses

Luca Russo, Luca Alinovi and Günter Hemrich

Abstract

This final chapter highlights the crosscutting issues that emerged from the six chapters based on case studies. It is organized around the issues of: existing limitations and challenges in responding to protracted crises; the critical role of institutions; and the effects on livelihood systems (adaptation and limits of resilience). It then reviews these cross-cutting issues in the light of challenges ahead. It proposes options and a related framework for looking at, and engaging with, crises differently.

Responding to protracted crises: Limitations and challenges

In the contexts of Sudan, DRC and Somalia discussed in this book, state and government presence has been almost non-existent, and international assistance has been the main – sometimes the only – source of public transfer for supporting immediate needs and providing a few essential services aimed at specific vulnerable groups.

In such contexts, interventions have focussed mostly on addressing the effects or immediate causes of food insecurity rather than their determinants: very rarely have international responses addressed the underlying and longer-term causes of food insecurity. This is due in part to the fact that humanitarian aid is virtually the only instrument available during conflict-related protracted crises and, as several authors rightly argue, its main focus should continue to be humanitarian (see Buchanan-Smith and Christoplos, 2004). But there is a gap in knowledge and a lack of adequate instruments for addressing the longer-term determinants of such crises – by supporting, for example, local formal and informal institutions, livelihood strategies and any positive adaptive mechanisms adopted by the population in the face of the crises.

Three overarching themes emerge from analysis of the responses described in these chapters: (1) short-term responses based on mainstream humanitarian (and sometimes development) paradigms dominate, while longer-term perspectives and alternative frameworks are not adequately explored; (2) linkages are generally weak between the overall policy context and the responses undertaken; and (3) governance issues are extremely important for delivering assistance.

Short-term responses and the mainstream paradigms

Responses to crises have been dominated by short-term interventions based on humanitarian principles (those of neutrality and saving lives in particular). They are generally undertaken outside state structures, with funding horizons of 6 to 12 months and high volatility of aid flows. There is a marked tendency to focus on food availability rather than on the access and stability dimensions of food security. The DRC and Jubba chapters pointed out the emphasis on agricultural recovery through distribution of free seeds and tools, while in the case of Sudan food aid represented nearly 60 per cent of all humanitarian assistance.

The chapters also included examples of weak links between short-term intervention time frames and longer-term issues. The distribution of free agricultural inputs in Jubba when local markets were functioning was likely to have a negative impact on those markets in the longer term. The free distribution of fishing equipment at Lake Edward in the DRC further exacerbated the fisherfolks' problem of overexploitation of fish and subsequent depletion of haelieutic reserves.

Alternatively, in protracted crisis contexts, mainstream longer-term programmes based on developmental paradigms (which generally include as objectives sustainability, participation and cost recovery) may not necessarily provide an alternative to humanitarian interventions. The agricultural support initiatives undertaken in the DRC and Somalia were focussed essentially on infrastructure rehabilitation and agricultural production and did not appear to offer long-term solutions to the structural and institutional causes of food insecurity. Development paradigm objectives have sometimes been applied uncritically, failing to take into account the context within which the activities were undertaken. In Somalia, for example, inadequate knowledge of clan politics led some humanitarian and development agencies, eager to promote participation, to work with local groups that represented militia factions rather than households and communities.

The chapters also illustrate how alternatives to mainstream approaches are possible. In Sudan, alternative response strategies were undertaken based on longer-term and local perspectives, even in the face of limited funding horizons. These strategies required reaching a common understanding and forging an alliance between all stakeholders through a buy-in process. NMPACT established a number of principles of engagement to be adhered to by all agencies. It promoted inclusion in the coordinating structure of the two warring factions through their humanitarian agencies, creating the basis for a sustained peace process at the local level. This approach marked a substantial shift away from OLS,[1] which had been based on the principles of 'neutrality' and 'impartiality', and so avoided entering into peace-building processes involving the warring factions. Within that framework, NMPACT promoted coordinated efforts based on priorities identified by the Nuba, which led to development of a common plan of interventions made up of a combination of short-term and long-term measures.

The pastoralist programme in southern Sudan aimed at addressing the vulnerability of pastoralists with multi-year interventions tackling those longer-term issues necessary for protecting and strengthening pastoralists' herds. It included capacity-building, cost recovery, involvement of local institutions and direct engagement with the knowledge and know-how of the beneficiaries (including pastoralists and traditional healers). This helped to strengthen local capacities, establish the basis for a sound and locally controlled policy framework and, to a certain degree, provide for the sustainability of the interventions.

In both of these cases, the interventions moved away from traditional mainstream emergency frameworks: they extended the use of humanitarian aid beyond mechanisms focused strictly on 'saving lives', and looked for ways to decrease vulnerability. Identifying, designing and implementing this type of intervention required complicated and difficult processes involving donors, local people and technical agencies. The process also required understanding multi-dimensional problems, complex and evolving contexts and longer-term strategic implications. The programme designs were characterized by the high degree of flexibility required by the evolving context, while they needed to respect both limited time frames and the scope for initiatives considered acceptable for humanitarian actions.

Weak linkages between political and policy environments and responses

Another cross-cutting theme emerging from the book is the weak linkage between political environments and the responses undertaken. There is a marked tendency on the part of the international community to ignore or downplay formal or informal policy processes as well as the overall political environment, whether because of concerns for humanitarian principles or lack of adequate in-depth analysis. This can have repercussions for long-term perspectives. In the contexts of protracted crises characterized by poor governance, formal policies are often of little relevance due to weak institutional and policy implementation capacities and to the fact that they often come from governments directly involved in the conflict. Therefore it is appropriate to use the broader definition of policy as 'a purposive course of action followed by an actor or a set of actors' (Anderson, 1994). The term can include formal donor and government policies, written declarations of intent or plans and more informal policy, which is rarely written down but is made apparent through decisions and actions and may be more relevant than any formal policy.

Underlying policies can be revealed in the humanitarian community's programming or in the responses of local institutions and communities in specific contexts. In Sudan during the 1998 Bahr el Gazhal crisis, clan and kinship structures refused to accept the targeting mechanism promoted by donors aimed at reaching those the outsiders perceived as most vulnerable (for example, households headed by women and IDPs) and instead used a redistribution mechanism within the community (Harrigin, 1998; Deng, 1999, quoted in Russo, 2005). These actions reflected a 'policy' on the part of local communities that perceived food aid as a common good to be used to strengthen long-term kinship ties and strengthen social capital over the longer term, rather than to address the short-term needs of a part of the community.

In Sudan there have been cases of warring factions utilizing food aid provided within a humanitarian framework. In some cases, the food aid was used to reach consensus and feed militias, but in others it was used to discriminate against 'hostile' groups by curtailing their access to assistance (Russo, 2005) – and thus more reflective of an underlying policy of war and conflict than of humanitarian principles. Also in Sudan, part of the humanitarian community in Sudan supported the ARS in the Nuba Mountains, in spite of the fact that it was instrumental to the GoS's policy of depopulating areas under SPLM/A control and a significant factor in the conflict.

The book includes examples of exceptions to the norm, where national and local stakeholders played important roles in defining the policy framework even in the absence of a formal state structure. The Nuba Mountains case study is particularly relevant in this respect. NMPACT included a policy-making structure that supported the coordination structure to orient collective decision making. The NMPACT Partners' Forum was a mechanism created to allow all implementing partners to meet systematically at a neutral location in the Nuba Mountains and was a crucial element in the strategy promoted by NMPACT to advance local ownership and conflict transformation.

Governance issues are of key importance in delivering assistance

In recent years significant attention has been paid to how to achieve improved operational coordination among parties and organizations seeking to provide assistance in protracted crises and humanitarian emergencies (Macrae, 2002). This book has identified coordination mechanisms as key factors of success and failure in the responses undertaken in various countries. The two DRC case studies point to the complete lack of coordination mechanisms as a key shortcoming of external interventions. In both Sudan and Somalia there were formal coordinating structures for international assistance, although OLS and the SACB had very different mandates.

The existence of a formal coordination mechanism is a condition that is necessary but not sufficient for improving the effectiveness of aid delivery. Coordination mechanisms need to be strengthened by agreed-upon enforcement mechanisms and should be based on shared basic operational principles, clearly spelt-out partnership arrangements and common frameworks for humanitarian and long-term interventions. The book provides illustrations concerning codes of conduct developed by stakeholders, especially in Sudan.

Another attribute of good practice in long-term development initiatives (and a governance issue) is the involvement of local partners and institutions in the definition and implementation of responses, yet such involvement is the exception rather than the rule in most interventions covered here. Local partners and governments are often excluded from the development of humanitarian responses because they could be perceived as (or actually are) a party to the conflict. In the DRC local partners were expected to simply execute projects defined by donors and were generally marginalized; initiatives with long-term perspectives and identified by local stakeholders as important (such as the land dispute committees) received very little external support. In Somalia there is a large number of Islamic civil society organizations with significant potential for providing access to services such as water, education, vocational training and health and for providing a popular political alternative to Somali clannism, violence and state collapse, but they are ignored by traditional donors.

In contrast, in Sudan there were systematic efforts to involve national and local stakeholders (including NGOs, pastoralist associations and government services) in establishing priorities and implementing responses, with important and positive consequences. NRRDO, a local NGO, helped international partners establish the rules of the game on the basis of Nuba priorities.

Local partners were an essential component of the more successful interventions to address longer-term needs in DRC, Somalia and Sudan, but the identification and involvement of local partners remains a highly debated issue among both humanitarian and development camps (see, for example, Longley and Maxwell, 2003; Slaymaker et al, 2005; HPG, 2006). It is particularly difficult to make decisions regarding local partner involvement in the absence of adequate analysis.

The critical role of institutions

The case studies provided ample evidence of the institutional changes caused by a general, prolonged lack of governance – which was often at the root of the conflict – and by conflict itself, and of the impact of these changes on livelihoods and the resilience of food systems. The resulting chapters highlight the crucial importance of the institutional context in mediating, for better or worse, the food security impacts.

We adopt a broad definition of the term 'institutions' to encompass processes that occur outside formal institutions but are particularly relevant to the context of fragile states in conflict situations (North, 1990): 'institutions are the rules of the game in a society, or more formally are the humanly-devised constraints that shape human interaction.' The case studies provide examples of how institutional dysfunctionality started well before the conflicts erupted, and how it contributed directly or indirectly to them and to food insecurity.

In the case of DRC and Somalia formal institutions were already undergoing a phase of deterioration prior to the conflicts and were unable to play a role in guaranteeing the access of certain groups to vital natural resources or basic services.

Access to land and institutional issues related to land tenure emerge from several studies as a crucial factor of conflict and food insecurity. In the case of DRC, land tenure issues had long been a major factor in poverty and discrimination. Various local wars then exacerbated the situation, and predatory behaviour in relation to land, and land itself, became factors in them, with politico-military elites consolidating their power bases and rewarding their supporters by extending control over land.

In Jubba (southern Somalia) as well, wars and patronage led to the dispossession of land, particularly that of marginalized and fragile groups such as the Bantu minority. In the case of Sudan, the (negative) role of formal institutions was important enough to be identified as a key triggering factor of the conflict. The GoS's policies affected the level of food insecurity in the Nuba Mountains through the Unregistered Land Act of 1970 (which set the stage for land-grabbing for mechanized farming schemes) and later through the displacement of Nuba people in 'peace village' policies and the blockade of humanitarian aid in SPLA-controlled areas. These measures disrupted traditional farming systems and had a severe impact on the Nuba people and Nuba agro-ecology.

The case studies also provide evidence and examples of how informal institutions such as social norms, kinship-based safety nets and community-based regulation of natural resource management are relevant to food security and how they have been particularly affected by the protracted crises. When the traditional regulatory agreement between Arab pastoralists and Nuba farmers in Sudan was disrupted by the conflict, it had direct effects on food security. In other parts of Sudan, breakages of social norms as a result of the conflict led to more destructive and ruthless cattle raids, with massive displacements and loss of lives and livelihoods. Both studies on pastoralism observed that the regulatory function exercised by local institutions over water and pasture were weakened by conflict, and this led to overexploitation of natural resources and had a negative impact on pastoralists' livelihoods.

Another element with a direct bearing on food security in case of massive and prolonged crises is the disruption of traditional kinship-based mechanisms that provide social protection. The 1998 conflict-driven famine in Bahr El Ghazal (Sudan) was identified by some Dinka groups as the *famine of breaking relationships* (Deng, 1999), which led to social entitlement failure and the supplanting of traditional elder authorities by military authorities. Similarly, in the Masisi region in eastern DRC, social structure and related kinship-based mechanisms for land distribution, already worn down prior to the conflict, eventually collapsed due to it. Many households were increasingly excluded from access to land and could no longer rely on the mechanisms of distribution and solidarity provided by the customary social structure.

Adapting institutions to the crisis

The chapters also provide evidence that the capacities of the local population and local institutions to adapt and eventually exploit the changing circumstances can mean that conflicts and institutional changes do not necessarily lead to negative food security outcomes for everyone. The absence of formal institutions and regulatory functions in eastern DRC favoured the movement of people from Lake Edward to the Virunga National Park. This offered fisherfolk who had become food insecure because of the depletion of fisheries resources the opportunity to create an agricultural-based livelihood for themselves.

Chapter 6 in this volume and other studies have shown how in Jubba, Somalia, local institutions adapted to the conflict; local markets continued to function during the conflict and were extremely relevant to food security. The study of Beni Lubero in the DRC challenged another commonly held assumption: it was found that farmers confronted with acute crisis situations do *not* always withdraw into subsistence farming. On the contrary, the revenue acquired from the sale of part of the agricultural production can constitute a central element of local food security.

Another remarkable example of adaptation was the capacity of Somalia to reconstitute a national monetary system within two years of the state's collapse in 1991. This achievement was very important for regions like Jubba because of its reliance on commerce. For rural consumers, an unstable currency can translate into local food shortages, high and rapidly shifting prices, and deteriorating terms of trade.

The effects on livelihood systems: The limits of resilience and adaptation

A feature distinguishing short-term shocks (such as floods or droughts) from protracted crises is the impact they have on people's livelihoods. While the impact of short-term shocks can be of a temporary nature and mitigated by people's coping strategies, in the case of protracted crises the effects tend to be of a more structural character. Unlike natural catastrophes, protracted crises are often characterized by conflicts as well as by an absence of public services, including security, health, education and regulations in the productive and trade sectors – all of which may lead to the sustained erosion of the livelihoods of specific groups, resulting in structural vulnerability.

All the case studies in this volume show that conflict and governance-related protracted crisis situations caused considerable erosion of household and community assets and led to substantial changes in the livelihoods basis of the affected populations. A number of cross-cutting issues and common trends emerged in relation to the resilience and adaptation capacity of livelihoods.

The book provides examples of several commonly occurring losses that affected people's key assets and that occurred due to lack of security that should have been provided by the state or by local institutions. One stark example is the substantial reduction or disappearance of livestock from people's livelihoods basis, as in Nuba in Sudan and Jubba in Somalia. Insecurity of land tenure provoked a change in cropping patterns in eastern DRC and the Nuba Mountains, while loss of labour opportunities occurred almost everywhere. Changing levels of access to markets, with a general worsening of the terms of trade for pastoralists and agro-pastoralists (as in Jubba) and also in some cases for farmers (as in eastern DRC) were also found to be common features of crises.

Most of the changes in livelihoods bases that occur in prolonged crisis situations are not short term. Typical traditional coping mechanisms, such as shifting of crop patterns or gathering of fruits and wild leaves, Davies (1993) defines as 'short-term, temporary responses to declining food entitlements, which are characteristic of structurally secure livelihood systems'. Instead, in the six areas studies for this book, the changes were of an adaptive nature, both in negative and positive terms, and showed that farmers and other vulnerable groups do have long-term visions of the crises and are in certain circumstance able to exploit the 'opportunities' offered by some crises. Farmers displaced by the conflict and afflicted by dwindling access to cultivated land due to population pressure moved from central to western Lubero, where conditions were better, at least for the time being. In Jubba, pastoralists and agro-pastoralists gradually moved toward agriculture as a normal response to population demands and volatile grain prices.

With respect to degrees of resilience, it was shown to depend to a large extent on the level of assets accumulated at the household level. According to evidence gathered from the FSAU data base, the non-poor ('better off')[2] are able to avoid poverty and hunger by selling off or consuming their assets and pursuing other strategies as the effects of a shock progress. These households are also likely to have strong social networks and relationships that they can draw on in times of need. The chronically poor and food insecure, by contrast, do not control enough assets to ever get ahead and in extreme cases they may become so poor that they can become destitute – a condition where a person cannot survive without sustained assistance (see Barrett and Carter, 2002; Sharp et al, 2003). The numbers of households in Somalia falling into the most vulnerable categories have substantially increased recently, which shows not only the negative effects of the crisis on people's assets but also the very limited effectiveness of international responses for offsetting them.

In the Somalia context pastoralist systems are proving to be more resilient than other systems. This resilience appears to be due to:

- *Mobility*. In spite of periodic constraints caused by insecurity, livestock-based systems have remained relatively mobile and herders have been able to practice their main risk-management strategy, which is *movement*.
- *Cross-border trade*. The area's proximity to primarily Kenyan but also Ethiopian markets has allowed local households to earn market income

from it, which has helped counter the loss of other sources of income in the past 15 years.

- *Robust markets*. These have generally functioned in large parts of the region, allowing pastoral households to weather volatile conditions by selling livestock products and helping them to restock herds after disasters.
- *Regional remittances*. These have helped pastoral households cope with income shortfalls and, in some cases, to invest in small business and other non-farm activities.

Civil society organizations have also made important contributions in supporting the adaptive capacities of vulnerable groups and strengthening resilience. In Walungu, DRC, the only organizations playing a role in land disputes (the main cause of conflict) were informal community-based *chambres de paix* (peace councils). Local Nuba organizations in Sudan discouraged negative short-term approaches such as unrestricted food aid distribution, which could have eroded the overall resilience of food systems. Kinship support also played an important role in strengthening resilience in the case of the Nuba. Transfers are also important factors in communities resilience. In Somalia, transfers of money from Somalis living outside the country (estimated at 22.5 per cent of GDP) have played a crucial role in the economy and in the protection of Somalia households.

While the considerable adaptive capacities of local population and institutions were a constant in the areas studied, there is concern for the *overall resilience* of the society and, in particular, of food systems – that is to say, 'the measure of a system to remain stable or to adapt to new situation without undergoing catastrophic changes in its basic functions' (Pingali et al, 2005). The agricultural production within Virunga National Park in DRC was soon affected by environmental degradation, and the alternative livelihoods strategy faced obstacles from the same 'institutional' factors from which the benefiting households (former fisherfolk) tried to escape – namely the 'environmental services' that were a factor in the destruction of Lake Edward's productive output. In Sudan, Nuba farmers also adapted their farming systems to the conflict situation by concentrating cultivation practices in the more secure hilly areas. However, such a system will likely have negative repercussions on the agro-ecology of the area, given that the traditional pre-conflict farming system comprised three separate pieces of land: the house farm, the hillside farm and the 'far' farm in the clay plains, cultivated with long-season sorghum and groundnuts.

The international community very often missed the opportunity of supporting the adaptive efforts of local communities. Opportunities created by civil society organizations were often ignored, if not undermined, while the role of remittances was also normally ignored or underplayed in needs assessments and vulnerability analyses undertaken by external actors. Given that social safety net mechanisms tend to break and lose effectiveness as crises progress or increase in intensity, analysis of their role should be attentive enough to take the stages of deterioration into account.

Rationale for a longer-term perspective: Informing responses and looking for conceptual, policy and operational frameworks

This book provides ample evidence of how the food security-related impacts of protracted crises differ from those of short emergencies, and how such crises call for different sets of responses. It illustrates how the protracted nature of the crises tends to have long-term negative impacts on people and institutions that can scarcely be addressed by mainstream short-term humanitarian and emergency responses. It also provides evidence of some conflicts as late manifestations of long-term institutional dysfunctioning and shows that in these cases the erosion of livelihoods had started well before the conflict erupted. The book chapters highlight the importance of institutions as a potential determinant of the crises but also as a potential factor of resilience. In many circumstances, people and organizations affected by the crises were able to organize themselves and adapt to the new environment irrespective of the external humanitarian assistance provided. Adaptation mechanisms followed patterns and modalities that external agencies usually did not perceive, understand or support. Furthermore assistance driven by exogenous programming mechanisms often undermined formal and informal institutions, local knowledge, traditional food systems and the local economy, thus weakening the adaptive capacity of communities and their capacities to recover or strengthen resilience.

The book makes evident the inherent fragility of the bases for people's livelihoods and food systems in contexts where government, private institutions and external agencies are unable to provide a minimum level of social protection and economic development. In fact, the key conceptual issue that planners face in contexts of protracted crises is how to address long-term problems in the absence of (or with failing) institutions, and where most interventions are going to be short term. The innovative interventions described in these chapters – and those with greater awareness of institutional aspects – seem to be an exception rather than the norm.

The contextual analyses presented by the chapters have drawn attention to two major challenges to increasing the relevance of food security-related response: the inadequacy of the analysis of context (institutions, policies, political economy), which limits the definition of the range of relevant responses; and the shortcomings of existing conceptual, policy and operational frameworks for dealing with food security in protracted crisis contexts.

Narrow analysis, narrow responses

The areas studied in this book show food insecurity to be on the whole a manifestation of social and political configuration. This reconfirms the essentially political nature of famine and food emergencies (see Sen, 1981; and de Waal, 1993) and the need to incorporate the institutional, policy and livelihoods dimension of the crises into food security responses (Devereux, 2000). Yet most mainstream analytical tools utilized, particularly in the DRC and Sudan, have treated food insecurity as triggered by natural hazards such as crop failure, or at best as livelihoods crises at the household level caused by external factors. Furthermore, most of the analysis undertaken has been for the most part sectoral and on the whole geared towards the identification of the basic needs of the affected populations, generally focusing on food deficits and key livelihood protection actions. It appears to have been geared towards identifying needs that correspond to the capacities of intervening agencies to deliver specific goods rather than to a contextual analysis. (See the examples of fishing equipment supplied at Lake Edward and seeds and veterinary medicines supplied free of charge in Somalia.)

The crises were characterized by institutional dysfunctioning or collapse, and the disruption or collapse of livelihoods, with an overall reduction in the society's resilience. In some cases, the interaction of institutional breakdown and conflict provoked the development of new, non-state centres of authority that consolidated themselves around alternative patterns of social control, protection and profit. Understanding such interactions requires a certain level of political-economic, policy and institutional analysis, which, when undertaken at all, tends to remain confined mostly to academic circles, with little impact on policy processes.

Mainstream analysis has contributed to the prioritization of investments dominated by humanitarian and emergency paradigms. Responses that would have helped address, and sometimes prevent, some of the determinants of food insecurity – such as land tenure insecurity, natural resource (mis)management, limited local capacities, insecurity of fragile groups – have represented only a tiny percentage of external assistance. Broadening the scope of analysis is challenging due to the absence of reliable and current information, the continuous and sudden changes of highly volatile contexts and response frameworks that are limited mostly to humanitarian responses; but doing so could help form the basis for a wider range of responses.

There are experiences in Somalia that represent a notable exception to the general trend. The FAO-supported FSAU for Somalia has gradually expanded the scope of its analysis: a forum approach encourages partners to share analysis and engage in building consensus, thus favouring interactive communication with decision-makers (Hemrich, 2005). The FSAU's recent development of the Integrated Food Security and Humanitarian Phase Classification (IPC) has provided an opportunity for shared analysis leading to commonly-agreed classifications. This in turn can provide the platform needed for a multi-dimensional analysis aimed at informing responses that address both the immediate and underlying causes of food insecurity (Flores and Andrews, 2007).

The lack of conducive operational, policy and analytical frameworks for intervening

The cases discussed point to how the unique character of the different contexts and the issues to be addressed call for a broad set of responses by the international community and national and local stakeholders. However, the range of response options has been very limited due to the existing aid delivery mechanisms and lack of conducive analytical as well as policy frameworks for intervening in protracted crisis contexts. Short-term interventions based on humanitarian paradigms have been dominant in all three countries and humanitarian aid, while indispensable for addressing the immediate needs of the most vulnerable population groups, has been inadequate for addressing structural causes of the crises. The principles of neutrality and impartiality may sometimes curtail the options for addressing the underlying political causes of the crises.

The development–relief continuum approach commonly employed by agencies is at odds with *neither peace nor war* situations where the transition from a crisis situation to peace and development is unlikely to be linear. At the same time, a rigid application of developmental paradigm elements such as sustainability also often fails to fit the protracted crisis contexts. The FAO twin-track approach and livelihoods-based frameworks provide interesting opportunities for bringing longer-term considerations into the picture in emergency contexts. However, they both have shortcomings. The twin-track approach is not structured to capture the institutional context, while livelihoods-based frameworks by being all-inclusive tend to underplay food security issues. Furthermore the operational implications of both approaches have yet to be fully defined (for example, how to use a livelihoods framework when government bodies are organized by sectors). In addition, lack of local ownership and the failure to capture the resilience dynamics of a society are cross-cutting limitations of all existing conceptual and operational frameworks.

Towards new conceptual frameworks

None of the existing conceptual and operational frameworks are fully satisfactory for addressing the challenges entailed by protracted crisis situations. The cases in this book are useful for identifying the methodological and operational challenges that need to be addressed and in providing a number of lessons for building future responses strategies. What we advocate here is the need to move beyond the existing conceptual frameworks to take into account the specificities of protracted crises contexts under a food security perspective.

Figure 11.1 shows a possible conceptual model for addressing food insecurity in protracted crisis contexts. It uses an overarching livelihoods-based framework to represent various processes that may occur at the global level and the effects of those processes at the household level. It also draws

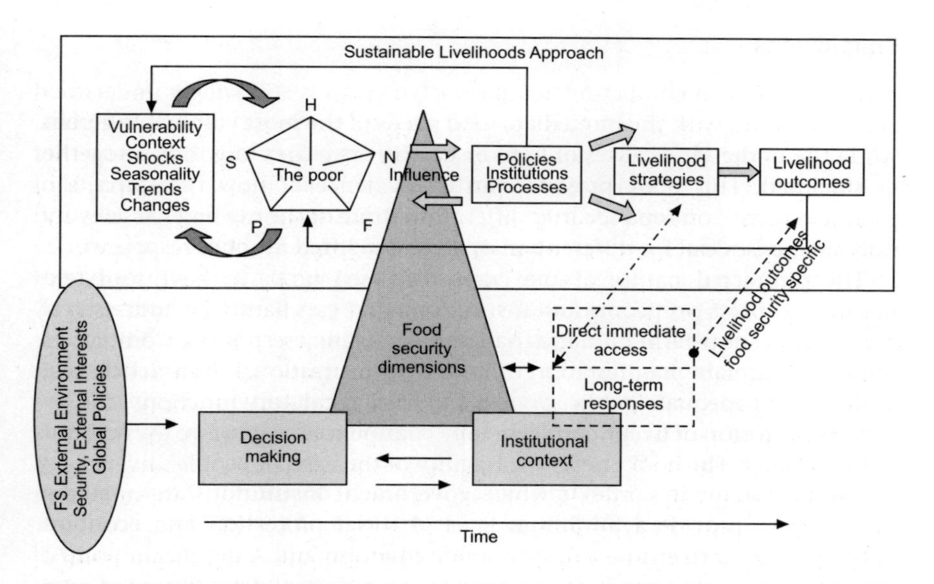

Figure 11.1 Conceptual model for understanding food insecurity in protracted crises through the sustainable livelihoods approach

attention to processes affecting livelihoods in a food security perspective and to the systems and processes of two domains that have emerged as driving factors for food security and insecurity in protracted crisis situations: decision making and the institutional context.

Analytical focus on the decision-making domain should contribute to the process of identifying and highlighting the responses (policies and programmes) that need to be made with respect to food security. In relation to the institutional context, it is important to identify those institutions that play a positive (or negative) role in strengthening the resilience of food systems, and those that are necessary for implementing food security responses. The model also illustrates the relationship between the two domains, which is not necessarily linear.

The proposed framework underlines the importance of the external environment for food security, particularly in a protracted crisis context where factors such as conflicts, security and external interests point to the need for political and economic analysis. The FAO twin-track approach can be very useful for making this conceptual framework operational. In particular, it can provide a temporal dimension to food security responses; and it can ensure adequate support to livelihood strategies through a mix of initiatives that ensures immediate access to food while addressing longer-term food insecurity determinants.

Conclusions

Food insecurity in conflict-related protracted crises is still widely understood as synonymous with the immediate food needs of the most vulnerable groups, while the underlying causes of food insecurity are either forgotten altogether or sidelined. This book provides further evidence of how the impacts of protracted crises on food security differ from those of shorter emergencies and how such crises call for different analyses and a different set of responses.

The protracted nature of the crises discussed tends to have long-term negative impacts on people and institutions that can hardly be addressed by mainstream short-term humanitarian and emergency responses. Conflicts are often the ultimate manifestation of long-term institutional dysfunction such as the lack of adequate public services and basic regulatory functions, and the process of erosion of livelihoods generally commences well before the outbreak of the conflict. The book shows the fragility of the basis of people's livelihoods and food systems in contexts where government institutions are unable or unwilling to provide a minimum level of social protection and economic development, or to ensure a proper policy environment. A significant point of the book is the importance of institutions as a potential determinant of crises but also as potential factors of resilience.

In many circumstances, people and organizations affected by crises have been able to organize themselves and adapt to the new environment irrespective of any external support provided. People and local institutions maintain their long-term perspectives even in very volatile contexts, and their adaptation mechanisms follow patterns and modalities that are generally not perceived, understood or supported by external agencies. Information gathered and analysis used in many crisis situations does not integrate the knowledge on local institution and policies necessary for supporting longer-term programmes (including peace processes) when appropriate.

Within a humanitarian framework, the fear of engaging with the institutional and policy contexts in order to avoid the risk of 'politicization' of the response can lead to the inadvertent promotion of interventions that fuel the very determinants of the conflict. Some of the innovative and institutionally sensitive interventions described in this book appeared to be the exception rather than the rule, and were the result of isolated initiatives rather than of systematic approaches.

There is an obvious lack of aid instruments and conceptual and operational frameworks to address food security in conflict-related protracted crises and fragile states. During conflict-related crises the humanitarian response framework continues to be virtually the only intervention mechanism, which often undermines those initiatives that require longer-term perspectives. On the one hand it remains indispensable to ensure neutrality for immediate responses that protect the most vulnerable. On the other hand, it is crucial to take into account institutional and policy dynamics that can support processes to rebuild resilience, create opportunities for strengthening the livelihoods of affected populations at the very early stages of the crisis and develop an adequate basket of interventions to address a variety of needs.

The current architecture of aid is changing to address the need for flexible funding in those countries where the political situation is evolving from conflict toward peace, in order to address the longer-term needs inherent in unstable political environments. Several donors have developed initiatives to promote peace processes in post-conflict situations (Lockhart, 2005).

The main conclusion of the research work in DRC, Somalia and Sudan is that the crises analysed by the six case studies had a long-term impact on food security and showed a multi-dimensional structure involving different temporal and causal dimensions. It is thus of paramount importance to understand and address simultaneously both immediate needs and the institutional, policy and livelihoods dimensions of crises – to decrease vulnerability while building viable and resilient mechanisms in the affected societies. This would require a rethinking of the aid delivery mechanisms and architecture, innovative and forward-looking approaches, strategic alliances and political will.

References

Anderson, J. (1994) *Public Policymaking: An Introduction,* 2nd edn, Houghton Mifflin Co., Boston.

Barrett, C. B. and Carter, M. R. (2002) 'Can't get ahead for falling behind: New directions for development policy to escape poverty and relief traps', *Choices* 16 (4): 35–38.

Buchanan-Smith, M. and Christopolos, I. (2004) 'Natural disasters amid complex political emergencies', *Humanitarian Exchange* 27, Humanitarian Practice Network, Overseas Development Institute, London.

Davies, S. (1993) 'Are coping strategies a cop out?', *IDS Bulletin* 24 (4): 60–72.

De Waal, A. (1993) 'War and famine in Africa', *IDS Bulletin* 24 (4): 33–40.

Deng, L. B. (1999) 'Famine in the Sudan: Causes, Preparedness, and Response', IDS Discussion Paper 369, Institute of Development Studies, Brighton.

Devereux, S. (2000) 'Famine in the twentieth century', IDS Working Paper 105, Institute of Development Studies, Brighton.

Flores, M. and Andrews, C. (2007) 'Linking analysis to response in fragile contexts: Emerging insights from the Integrated Food Security and Humanitarian Phase Classification (IPC)', presented at conference 'Fragile states – fragile groups: tackling economic and social vulnerability', 15–16 June 2007, World Institute for Development Economics Research (WIDER), Helsinki.

Harrigin, S. (1998) *The Southern Sudan Vulnerability Study*, Save the Children Fund (UK), Southern Sudan Programme, Jubba.

Hemrich, G. (2005) 'Matching food security analysis to context: The experience of the Somalia Food Security Assessment Unit', *Disasters* 29 (S1): S67–S91.

HPG (Humanitarian Policy Group) (2006) 'Providing aid in insecure environments: Trends in policy and operation', HPG Briefing Paper 24, Overseas Development Institute, London.

Lockhart, C. (2005) 'From aid effectiveness to development effectiveness: Strategy and policy coherence in fragile states', Overseas Development Institute, London, http://www.oecd.org/dataoecd/33/11/34258843.pdf.

Longley, C. and Maxwell, D. (2003) 'Livelihoods, chronic conflict and humanitarian response: A synthesis of current practice', ODI Working Paper 182, Overseas Development Institute, London.

Macrae, J. (2002) 'The new humanitarianisms: A review of trends', *HPG Report* 11, Overseas Development Institute, London.

North, D. (1990) *Institutions, Institutional Change and Economic Performance*, Cambridge University Press, Cambridge.

Pingali, P., Alinovi, L. and Sutton, J. (2005) 'Food security in complex emergencies: Enhancing food system resilience', *Disasters* 29 (s1): S5–S24.

Russo, L. (2005) 'Food security and agricultural rehabilitation with a medium- to longer-term perspective in the protracted crisis context of Sudan' (mimeo).

Sen, A. (1981) *Poverty and Famines: An Essay on Entitlement and Deprivation*, Clarendon Press, Oxford.

Sharp, K., Devereux, S. and Amare, Y. (2003) 'Destitution in Ethiopia's northeastern highlands (Amhara National Regional State)', Institute of Development Studies, Brighton and Save the Children (UK) Ethiopia, Addis Ababa.

Slaymaker, T., Christiansen, K. and Hemming, I. (2005) 'Community-based approaches and service delivery: Issues and options in difficult environments and partnerships', DFID, London.

Notes

Chapter 1

1 Several donors have developed strategies based on definitions that differ significantly from one another. For some, the characteristic of such states are weak policies and institutions for delivering services, controlling corruption or providing sufficient voice and accountability, with the risk of conflict and political instability (World Bank, 2005). Others emphasize inability or unwillingness to provide physical security, political institutions, economic management and social services (OECD, 2006) or failure to develop and implement pro-poor policies (OECD/DAC, 2005).

Chapter 2

1 Deng (2002) makes a distinction between endogenous and exogenous warfare because of the different impacts on social capital. Deng uses 'endogenous' to indicate conflicts that explode within the same community or ethnic group.
2 The average per capita GDP in Sudan was $395 in 2001 (World Bank, 2003).
3 In the northern sector only 256,000 beneficiaries (or 15 per cent of the caseload) were classified as war-affected.
4 Part of the food aid assistance has reportedly been utilized to feed soldiers of the warring factions.
5 In the Nuba Mountains, the traditional farming system comprised three pieces of land: the house plot (generally cultivated by women), the hillside plot and the 'far' plot in the clay plains, cultivated with long-season sorghum and groundnuts.
6 For instance, the irrigated and semi-mechanized sectors absorbed 99 per cent of rural credit.
7 The mechanism by which WFP, in response to a request from a government or the UN Secretary-General, provides emergency food aid and related assistance to meet the food needs of people affected by a disaster or other emergency. An EMOP generally lasts no more than 24 months, including extensions, after which any need for continuing relief and rehabilitation or recovery assistance is normally met through a protracted relief and recovery operation (PRRO).
8 USAID assistance to Sudan during the conflict was based on three principles: (1) the 'war on terrorism', with GoS identified as a sponsor of terrorism;

(2) support to the peace process to guarantee human rights and religious tolerance; and (3) humanitarian access.

9 The inputs-transfer component of the seeds and tools programme is estimated at 70 to 75 per cent, while in the case of food aid assistance that component is well over 80 per cent.

10 Since late 2005, the various SPLM Secretariats have been transformed into Ministries and have incorporated staff from the former GoS.

Chapter 3

1 In early 2002 representatives from the GoS and the SPLM and SPLA agreed on a total population estimate in the region of 1.2 million, with approximately 350,000 of these living in the SPLM/A areas at the time (Office of the UNR/HC, 2002f). However, NMPACT partners later that year estimated that the overall population figure was more likely to be close to 1.4 million, taking into account population return after the signing of the Cease-fire Agreement in January 2002 (Office of the UNR/HC, 2002f).

2 The MFC was replaced in 1995 by the Department of Rainfed Agriculture in the Ministry of Agriculture.

3 Later on the blockade was extended to SPLM controlled areas in southern Blue Nile and to National Democratic Alliance controlled areas in Eastern Sudan. In both cases, though, the population affected by the blockade was considerably smaller than that of the Nuba Mountains and in both areas people had the possibility of moving cross-border to other countries to obtain supplies, something that was not possible in the Nuba Mountains. Several international agencies were also able to set up logistical bases in border countries to deliver unauthorized aid to the affected populations in the two regions.

4 The information has been derived from a series of NMPACT documents, chiefly the information tables produced between 2002 and 2004, and from a stocktaking exercise detailing agencies' activities in South Kordofan state, which was prepared during the development of the Nuba Mountains Programme, NMPACT's precursor. The information tables can be found in Pantuliano (2005).

5 Many Nuba expressed their disappointment with the Two Areas Protocol as they felt it did not adequately address the fundamental grievances of Nuba communities that had led to conflict, particularly in terms of sharing of power and wealth and pursuit of justice. Many individuals and communities especially resented the change of boundaries, the new administrative structure of the enlarged State of South Kordofan and the failure to secure the return to the name 'Nuba Mountains' as an administrative title.

Chapter 4

1 These NGOs were: ACORD, ACROSS, ADRA, CDOT, NPA, Oxfam-GB, SC-UK, VSF-B, VSF-CH, VSF-G, Vetwork Services Trust and World Relief.

Chapter 5

1 For a good overview of politics in the Jubba region, see Menkhaus (1999; 2003).
2 This discussion excludes Somaliland in the northwest, which has its own currency and monetary system.

Chapter 6

1 This includes the Lower Jubba and Middle Jubba Regions, which stretch from the Jubba River to the east Kenya to the west, and the Indian Ocean to the south.
2 The Shebelle River of southern Somalia is often classified as a perennial river, but the lower parts of the river actually dry up during the dry season. Unlike the Jubba, the Shebelle does not exit to the Indian Ocean but instead disappears into swamps east of the Jubba River about 30 km from the coast.
3 For a good overview of politics in the Jubba region, see Menkhaus (1999; 2003).
4 *Deschek* farming is a type of flood-based irrigation that relies on the seasonal flooding of the Jubba River. It is found in large expanses in the middle and lower Jubba areas and can cover as much as several thousand hectares in certain locations.
5 My thinking about assets and the model presented in this paper were heavily influenced by my collaboration with Michael Carter of the University of Wisconsin, who has done some of the pioneering work on asset-based approaches to poverty analysis (see Carter and Barrett, 2006).
6 As used here, a TLU (tropical livestock unit) is:
1.0 TLU = 1 head of cattle (oxen, bull, cow, calf, heifer)
0.5 TLU = 1 horse/donkey/mule
1.4 TLU = 1 camel
0.1 TLU = 1 sheep/goat
0.05 TLU = 1 chicken
The TLU ratios approximate weight, subsistence (food) and market value of different animals.
7 I have taken some liberties with the FSAU herd range data to make it comparable to my 1987–1988 data. To make the comparison between the two periods even possible, I equated the 'better-off' group to my wealth quartile I; the 'middle' group to my wealth quartiles II/III; and FSAU's 'poor/very poor' group to my poorest wealth quartile (IV). Because the FSAU herd data are in ranges rather than averages, I took the upper end of the range to compare with the average livestock holdings for 1988. The comparative data should be treated as illustrative and not as exact measurements of the differences.
8 The distribution was completed within four weeks and reached more than 95 per cent of the targeted households. A study of the project found that (1) the bulk of the cash was used for immediate food consumption and needs, which helped to stimulate local trade and business activity; (2) there were very few cases of cash abuse; (3) cash grants empowered women by allowing them greater influence in household decisions; and

(4) cash transfers were more cost-effective than food aid distribution (Ali et al, 2005). In spite of the advantages it had, Emergency Cash Relief Programme needed to better integrate its activities with other investments to maximize their development impacts (Ali et al, 2005).

Chapter 7

1 This chapter is based on original research carried out by Tim Leyland, Abdi Osman Haji-Abdi, Andy Catley and Habiba Sheikh Hassan and presented in their report *Livestock, Markets and Food Security in Southern and Central Somalia,* written for FAO in 2006.
2 The use and interpretation of TLU/AAME ratios varies, and the calculation of the ratio is hindered by inaccurate human or livestock population figures in most Somali areas. Sandford and Habtu (2000) suggested that pastoral households with less than 3.0 TLU/AAME were highly vulnerable and Lybbert et al (2004) suggested a minimum 4.5 TLU/AAME if pastoralists were to avoid poverty.
3 The *dia*-paying group is also relevant in terms of the lifelong membership of its individuals despite their physical location. Consequently, a Somali living in the US, Canada or Europe will not only know his *dia*-paying group but will also be socially obliged to contribute to *dia* payments related to events in rural Somalia. This helps explain the strong global Somali communications network and the effectiveness of international money transfers by Somalis.
4 COOPI is an Italian NGO; EPAG is a British NGO.
5 There is no consensus on what to call these Islamic organizations; the phrase 'Islamic charity' is commonly used, but the English labels 'humanitarian' and 'charity' have no exact Arabic translations (Le Sage and Menkhaus, 2004).

Chapter 9

1 This study was prepared in collaboration with two assistant consultants and a team of students and former students from Butembo's Université Catholique du Graben.
2 The chapter highlights the role of institutions as determinants of livelihoods resilience, which is in line with other theoretical and empirical findings regarding livelihoods adaptation to protracted crises (see, among others, Ellis (2000) and the presentations of the 'FAO International Workshop on Food Security in Complex Emergencies: Building Policy Frameworks to Address Longer-Term Programming Challenges', held in Tivoli, Italy, 23–25 September 2003.) This is not to say that other factors (such as shocks) do not play a role in livelihood resilience during protracted crises. But institutional issues are important in considering appropriate responses to food insecurity because they determine livelihood access to vital assets – a problem that is important especially in relation to the management of scarce resources during post-conflict situations (see Ballentine and Nitzschke (2005), among others).

3 This starting point differs fundamentally from Hyden (1980), who claims that in times of crisis, households return to subsistence farming as a prinicipal method of survival. While this might be true in part, increasing evidence suggests that other coping strategies are as important in ensuring livelihoods survival during such situations, such as selling or buying food on markets (see Collinson, 2003; Levine and Chastre, 2004). A particularly interesting analysis of a Congolese case study in this regard has been provide by Pottier and Fairhead (1991) and Fairhead (1992).

4 In greater detail, the factors identified as influencing agricultural production were: climate (dryness, excessive rain); environment (erosion, plant diseases); demography (population pressure, lack of family planning); access problems (distance and access to arable land); and institutional factors (lack of a state framework, political interference). Those identified as influencing agricultural revenue were: price (i.e. what price for what product, what price in which season); quantity put on sale; organization of the market (i.e. direct sale or sale through intermediaries/cooperatives); taxes (including harassments by the military and other 'negative forces'); means of transport; means of conservation; and transformation.

5 The northern parts of North Kivu are inhabited almost exclusively by the Nande, a Bantu group that has its origins in the ancient kingdom of Kitara (Remotti, 1993). The self-designation of the Nande is 'Yira' or 'Bayira'. More than a global ethnic designation, this term refers to a social stratum of agriculturalists that were opposed to the pastoralist ruling class of the Bahima. (This division is also observed in other traditional interlacustrian kingdoms such as the Toro, Ankole and Bunyore.) The Zairian part of the Bayira community can be divided into the following sub-groups or '*chefferies*': Banyisanza Bashu, Baswagha, Batangi and Bamate.

6 In contrast, nearly 90 per cent of the local population had a concession insufficient to fulfil even their daily needs.

7 This dynamic reflects an important difference with, for example, Masisi, where agricultural profits were invested foremost in cattle (Vlassenroot, 2002).

8 For a discussion of artisanal gold exploitation in North and South Kivu during the 1980s, see MacGaffey (1987).

9 Especially after Mobutu's liberalization of the Zairian mining sector in 1982–1983, a new generation of transborder traders started to smuggle gold from the Kilo Moto mines in Ituri to buyers in Uganda and Burundi. This gold was in turn exchanged for household and other products from Asia and East Africa, or for dollars, which were again injected into the agricultural bulk trade. In a way, gold thus saved this region from the rampant inflation that was felt more acutely in other regions such as Masisi.

10 This gender dimension runs parallel with Masisi and South Kivu, where women traditionally carry food crops to the markets. Thanks to growing price fluctuations (due to numerous factors such as regular drought, different cropping seasons at low, middle and high altitudes, but also intervention by intermediary bulk traders), the opportunities for this food trade grew during the 1980s and 1990s, which simultaneously gave women

a better nutritional and economic (power) status within the household (Newbury, 1984; Pottier and Fairhead, 1991).

11 The Rassemblement Congolais pour la Démocratie was founded in 1998 in Rwanda as a reaction against the Laurent-Désiré Kabila regime. It was originally presided by Professor Wamba-Dia-Wamba. After the professor's replacement by Emile Ilunga, the movement split into two factions, the RCD-ML (backed by Uganda) and RCD-Goma (backed by Rwanda).

12 The Mouvement pour la Libération du Congo and RCD-National were founded in 1999/2000 in Uganda in an attempt to open another Uganda-backed front against Laurent Kabila. The MLC leader is the warlord and businessman Jean-Pierre Bemba, while RCD-N is headed by Roger Lumbala.

13 The Union de Patriotes Congolais was founded in 1999 by the Gegere (northern Hema) Thomas Lubanga, and sustained consecutively by Uganda and Rwanda. Its leader is currently being probed for war crimes in The Hague. (For a detailed discussion, see Vlassenroot and Raeymaekers, 2004b.)

14 For a more detailed description of this relationship, see Chapter 2 in Vlassenroot and Raeymaekers (2004a).

15 In May 2005, southern Lubero hosted 272,240 IDPs, of a total of 685,000 in the whole of North Kivu (OCHA, 2005).

16 According to OCHA (2005), the limited importance of food aid in this context should not lead to an underestimation of the impact of food aid in general, but could (paradoxically) indicate a positive perception of its effectiveness, especially in regions devastated by prolonged warfare.

17 The Forces Armées de la République Démocratique du Congo is Congo's newly-unified army.

18 In May 2005, southern Lubero hosted 272,240 IDPs, of a total of 685,000 in the whole of North Kivu (OCHA, May 2005). In December 2006, another 7,000–10,000 people reportedly crossed the border at Rutshuru into Uganda, following armed clashes there between the FARDC and dissident brigades (OCHA website).

19 Livelihood 'diversification' is frequently considered a factor that helps explain the resilience of households in situations of food insecurity. According to Ellis (2000), rural diversification involves 'the process by which rural households construct an increasingly diverse portfolio of activities and assets in order to survive and to improve their standard of living'.

20 The local food basket in North Kivu is made up of subsistence farming (53.7 per cent), sale (27.6 per cent), barter (2.3 per cent), seeds (11.4 per cent), and clothing and various (5 per cent). Average household spending in the province is for: food (39 per cent), medical care (24.2 per cent), education (27.5 per cent) and other (9.3 per cent – including clothes, soap, investments, etc.) (OCHA, 2005).

21 Indeed, an important condition for households to consume their seeds and/or genitors as a response to acute food crises is that they actually *have* a subsistence stock; this probably helps explain why this strategy figures more prominently in Beni than in Lubero (see Table 9.1).

22 Estimates were made on the basis of discussions with farmers from central and southern Lubero concerning their principal sold products.
23 Woolcock (1998) explains this problem as follows. Where generalized trust (or 'the disinterested cooperation of many individuals' (Einstein)) extends only to members of the same community, a stark non-developmental reality is likely to be present. Such situations are often characterized by an 'excess of community', built on fierce ethnic ties and familial loyalties that result in a lack of both economic and social mobility as well as the impossibility of amicably resolving disputes with outsiders. This idea comes close to what Granovetter (1973) calls the strength of 'weak ties', which enable people to develop economically *outside* their small community. In a situation characterized by strong horizontal ties, therefore, the breaking point that enables communities and groups to overcome the problem of underdevelopment is the existence of vertical ties social ties, namely of relationships outside the immediate community or group.
24 Traditionally, two normative systems regulate the right to local land tenure in Beni-Lubero: while private land ownership is applied to most urban areas as well as to certain large concessions such as plantations and mines, the access of smallholders is usually regulated by customary law. Small farmers can secure their rights to an arable field through the payment of an annual or seasonal tribute to the local customary chief.
25 Conversation with the author, December 2005.
26 In central Lubero for example, nearly three-quarters of intervening agencies (73 per cent) focus narrowly on the aspect of food production in the form of distribution of seeds and tools. The preponderance of interventions focusing on food availability rather than sustainability is also reflected in the number of organizations involved in each: while nine are active in availability (Oxfam, *Solidarités,* World Vision, CESVI, *Première Urgence,* MSF, Norwegian Refugee Council, Save the Children, SODERU), only three are active in sustainability (AAA, FAO, VECO RDC).
27 For example, in the area of Isale, where the FARDC is combating Ugandan militias, the Congolese army is currently imposing a system of 'unfree labour' on resident civilians. Many of the region's IDPs are reluctant to return home because they fear they will have to work for the Congolese soldiers and give them food: 'If it were not for the abuses by the soldiers, we would have returned home by now', one resident of North Kivu told the UN news network in May 2006 (IRIN, 26 May 2006).
28 Exceptions to this rule are VECO's food security assessments in the areas of Bulambo, Bunyuka and Kyavinyingo, and – to a lesser extent – the programme overview of Agro-Action Allemande. Other agencies continue to focus on the needs of IDPs, particularly in the north and south.
29 FAO describes food security as comprising four dimensions: food availability, food access, stability and food utilization. The twin-track approach builds on these dimensions of food security by combining investments in agricultural development/productivity enhancement with targeted programmes to enhance direct and immediate access to food for the most seriously undernourished. For a deeper discussion, see Pingali et al (2005).

Chapter 10

1 In this chapter, a food system refers to the different activities that guarantee community's food security, including the production, distribution and consumption of food. Particular attention will be given in this analysis on the institutional variables defining people's production of, access to and consumption of food.
2 These case studies were based on data collection exercises involving interviews with key informants and group discussions with village members. The group discussions involved two different exercises. The first aimed at acquiring a better understanding of the shifts in local patterns of land use and land distribution. The second focused on household strategies and the impact of food security interventions. Participants were asked to identify the main constraints to their food production and access to markets, to discuss the different short-term and long-term strategies that were developed to deal with the negative effects of land scarcity and insecurity, and to evaluate the impact of external interventions.
3 Land owned by churches was in most cases accessible only to members of the church. If contract conditions were not met, farmers were readily evicted from rented plots.
4 In certain areas, this legal parallelism even led to a resurgence of violent conflict: in the region of Masereka, entire villages were burned at regular intervals by angry farmers who had been chased from their land (often by their own family members) for unjustified reasons.
5 Given the general impoverishment of the population, even with the re-establishment of security, it is doubtful that cattle would return soon to Walungu. The effects of the quasi-disappearance of cattle should not be underestimated: these conditions risk preventing long-term economic recovery and the restoration of local social cohesion.
6 For a description of how farmers dealt with decreasing land availability during the 1980s in Bwisha, see Pottier and Fairhead (1991).
7 This figure is based on author interviews with food economy specialists and observations in the field.
8 Witchcraft was often used to prevent others from trespassing.
9 The Forces Démocratiques pour la Libération du Rwanda (FDLR) controlled large parts of the interior of North and South Kivu. The origins of the movement go back to Interahamwe militias involved in the execution of the Rwandan genocide in 1994. After their defeat, most of their leaders moved to DRC, from where they reorganized militarily and politically. During the Congolese war, they were allied to the Kinshasa government and the local rural Mayi-Mayi militias. Following the start of the peace process in July 2003, the FDLR was one of the main sources of insecurity in eastern DRC.
10 Interview with a local NGO leader, Walungu, December 2005.
11 This section is based on interviews with representatives from local and international agencies that developed food security interventions. These interviews were conducted in Goma, Walungu, Bukavu, Masisi and Lubero in December 2005. The food security interventions that were analysed, were those developed since the outbreak of the war. Even if some similarities

(for example, the use of 'short-term frameworks' and the neglect of the high level of locally produced and marketable food stuffs) can be drawn with food aid practices of international agencies when they worked in the Rwandan refugee camps based in Zaire between 1994 and 1996, these interventions are not part of the assessment.

12 Local farmers complained about the high interest rates they had to pay for their seeds (20 per cent regardless of productivity).

13 For example in central Masisi, imported maize distributed to local farmers by a humanitarian agency negatively affected the producers of local maize in nearby Rutshuru, who could not compete with the imported maize on local markets. In addition, most food for work initiatives were not preceded by assessments about the availability of labour and the need for food (interviews with local food security consultants, Goma, December 2005).

Chapter 11

1 OLS was a coordinating mechanism that included all the major UN agencies (UNICEF, WFP, FAO, OCHA and WHO) and some NGOs. OLS was the first humanitarian programme in the world established through a tripartite agreement (between the GoS, the SPLM and the UN) inside a foreign country to provide relief to war-affected and IDPs. It was established in 1989 following a devastating famine.

2 The wealth categories in FSAU's livelihood classification system are based on the notion of asset ownership in terms of livestock numbers, farm land size or a combination of the two. In the model, households are categorized according to initial asset holdings; for pastoral communities in the Horn of Africa the number for the 'better off' tends to be four or five TLU per capita.

Index